SpringerBriefs in Quantitative Finance

More information about this series at http://www.springer.com/series/8784

Anna Aksamit · Monique Jeanblanc

Enlargement of Filtration with Finance in View

Springer

Anna Aksamit
School of Mathematics and Statistics
University of Sydney
Sydney
Australia

Monique Jeanblanc
Laboratoire de Mathématiques et Modélisation
Université d'Evry Val D'Essonne
Évry
France

ISSN 2192-7006 ISSN 2192-7014 (electronic)
SpringerBriefs in Quantitative Finance
ISBN 978-3-319-41254-2 ISBN 978-3-319-41255-9 (eBook)
https://doi.org/10.1007/978-3-319-41255-9

Library of Congress Control Number: 2017954282

Mathematics Subject Classification (2010): 60-02, 60G07, 60G40, 60G44, 60H99, 60J65, 91B44, 91G40
JEL codes: G1, G10, G3, G32, G33, D8, D82

Printed on acid-free paper

This Springer imprint is published by Springer Nature
The registered company is Springer International Publishing AG
The registered company address is: Gewerbestrasse 11, 6330 Cham, Switzerland

Contents

Introduction

At the end of the 1970s, Jean Jacod, Thierry Jeulin and Marc Yor started a systematic study of enlargement of filtration which focuses on the properties of stochastic processes under a change of filtration. The main challenge is to find conditions under which, for two filtrations $\mathbb{F} := (\mathscr{F}_t)_{t\geq 0}$ and $\mathbb{G} := (\mathscr{G}_t)_{t\geq 0}$ satisfying $\mathscr{F}_t \subset \mathscr{G}_t$ for each $t \geq 0$, any $\mathbb{F}$-martingale remains a $\mathbb{G}$-semimartingale and, if this is the case, to give the decomposition of $\mathbb{F}$-martingales as $\mathbb{G}$-semimartingales.

Usually, only enlargements of filtration, as opposed to a generic change of filtration, are considered, in order to preserve adaptedness of stochastic processes. There are mainly two kinds of enlargement of filtration studied in the literature:

- Initial enlargement of filtration, that is the case where $\mathbb{G}$ is the smallest right-continuous filtration containing $\mathbb{F}$ and $\sigma(\zeta)$, denoted by $\mathbb{G} := \mathbb{F}\nabla\sigma(\zeta)$, where ζ is a random variable.
- Progressive enlargement of filtration, that is the case where $\mathbb{G}$ is the smallest right-continuous filtration containing $\mathbb{F}$ and $\mathbb{A}$, denoted by $\mathbb{G} := \mathbb{F}\nabla\mathbb{A}$, where $\mathbb{A} := (\mathscr{A}_t, t \geq 0)$ is the natural filtration of the process $A := \mathbb{1}_{[\![\tau,\infty[\![}$ and τ is a random time.

Up until now, three lecture notes have been dedicated to enlargement of filtration: Jeulin [137], Jeulin and Yor [142] and Mansuy and Yor [172]. The first two are not available in English. These volumes, as well as the book of Jacod [121], present deep results and contain many examples. One of the goals of our work is to recall some of these results. Chapter 20 of Dellacherie et al. [75] contains a very general presentation of enlargement of filtration theory, based on fundamental results of the theory of stochastic processes, developed in the previous chapters and books by the same authors. Chapter 12 in Yor [218] and the book Mansuy and Yor [172] are focusing on the case where all martingales in the reference filtration $\mathbb{F}$ are continuous. A survey containing many exercises can be found in Mallein and Yor [169, Chapter 10]. Protter [188] and Jeanblanc et al. [135] have devoted a chapter of their books to the subject. The lecture by Song [198] contains a general study of the subject. The reader can also find a summary and many examples of some classical problems in the lecture by Ouwehand [187]. The book of Hillairet and Jiao [114] contains applications to portfolio optimization.

Since its introduction in the late 1970s, enlargement of filtration has remained an important tool and field of study in the theory of stochastic processes. The theory has seen its second wind recently with revised interest sparked by applications in mathematical finance. These in particular include credit risk and modelling of asymmetry of information, where one considers a financial market where different agents have different levels of information. Results from the theory of enlargement of filtration are extensively used to study specific problems that occur in insider trading, in particular the existence of arbitrage strategies adapted to the large filtration $\mathbb{G}$. The theory of enlargement of filtration is also a cornerstone for the study of default risk.

At the beginning of each chapter of this book, we give the framework used in that chapter, even if this may lead to repetitions. Each chapter is concluded by bibliographic notes containing references to important papers and books. These lists are not exhaustive; nevertheless, we hope that we provide enough information for results or proofs which are not presented in this volume. Indexes of notation and keywords are given at the end of the volume. Exercises are given at the end of each chapter. The solutions and additional exercises can be downloaded from http://extras.springer.com.

The first version of these lecture notes was written to support the course given by Monique Jeanblanc and Giorgia Callegaro at University El Manar, Tunis, in March 2010 and later by Monique Jeanblanc at various schools: Jena, in June 2010; Beijing, for the Sino-French Summer Institute, in June 2011; CREMMA school, ENIT, Tunis, in March 2014; Moscow University, in April 2014; and as a main Master 2 course, Marne La Vallée, Winter 2011. Monique Jeanblanc would like to thank all the students for their participation.

We warmly thank Samuel Cohen, Claudio Fontana, Jean Jacod, Marek Rutkowski, Thorsten Schmidt and two anonymous referees for reading a preliminary version of this book and suggesting many improvements. Our work was supported by Chaire Marchés en Mutation (Fédération Bancaire Française), Labex ANR 11-LABX-0019.

Chapter 1
Stochastic Processes

We recall classical facts from the theory of stochastic processes, which will play an important role in the following chapters. We pay special attention to projections and dual projections which are essential for the study of random times. References for the recalled results are given in bibliographic notes at the end of the chapter.

1.1 Background

Let $(\Omega, \mathscr{F}, \mathbb{P})$ be a probability space and $\mathbb{F} := (\mathscr{F}_t, t \geq 0)$ be a given filtration on that space satisfying the **usual conditions** – $\mathbb{F}$ is right-continuous, i.e., $\mathscr{F}_t = \cap_{s>t}\mathscr{F}_s$, and the σ-field $\mathscr{F}_0$ contains all $\mathbb{P}$-null sets – and $\mathscr{F}_\infty := \vee_{t\geq 0}\mathscr{F}_t \subset \mathscr{F}$.

Throughout the book we shall always denote by *blackboard bold capital* letters the studied filtrations, e.g., $\mathbb{F}$, $\mathbb{G}$, $\mathbb{H}$, and by the corresponding *calligraphic capital* letters the σ-fields belonging to a given filtration indexed by $t \in [0, \infty)$, e.g., $\mathscr{F}_t$, $\mathscr{G}_t$, $\mathscr{H}_t$. Unless stated otherwise, any general filtration on $(\Omega, \mathscr{F}, \mathbb{P})$ considered in this book, is assumed to satisfy the usual conditions. As usual, for two σ-fields $\mathscr{G}$ and $\mathscr{H}$, we denote by $\mathscr{G} \vee \mathscr{H}$ the smallest σ-field containing all sets in $\mathscr{G}$ and $\mathscr{H}$. Similarly, for two filtrations $\mathbb{F}$ and $\mathbb{H}$, we denote by $\mathbb{F} \vee \mathbb{H}$ the filtration $(\mathscr{F}_t \vee \mathscr{H}_t, t \geq 0)$. Moreover, we shall write $\mathbb{F} \triangledown \mathbb{H}$, which is not a standard notation, to denote the filtration $(\bigcap_{s>t} \mathscr{F}_s \vee \mathscr{H}_s, t \geq 0)$. Note that $\mathbb{F} \triangledown \mathbb{H}$ is simply the **right-continuous regularization** of $\mathbb{F} \vee \mathbb{H}$.

Measurability of stochastic processes is treated in the following definition.

Definition 1.1 (a) A **stochastic process** (or just a **process**) is an $\mathscr{F} \otimes \mathscr{B}(\mathbb{R}^+)$-measurable mapping defined on $\Omega \times \mathbb{R}^+$ with values in $\mathbb{R}^d$. In particular, each t-section, X_t is a random variable.
(b) A process X is **progressively measurable** if for any $t \geq 0$, X restricted to $\Omega \times [0, t]$ is $\mathscr{F}_t \otimes \mathscr{B}([0, t])$-measurable.

A. Aksamit and M. Jeanblanc, *Enlargement of Filtration with Finance in View*, SpringerBriefs in Quantitative Finance,
https://doi.org/10.1007/978-3-319-41255-9_1

(c) A process X is $\mathbb{F}$-**adapted** if for any $t \geq 0$, the random variable X_t is $\mathscr{F}_t$-measurable.
(d) The **natural filtration** $\mathbb{F}^X$ of a process X is the smallest filtration $\mathbb{F}$ which satisfies the usual conditions and such that X is $\mathbb{F}$-adapted.

Remark 1.2 One might think that a collection $(X_t, t \geq 0)$ of independent r.v.'s may be chosen in such a way that the process X satisfies the measurability assumption. This is not the case: let $(X_t, t \geq 0)$ be independent centered r.v.'s and $\sup_t \mathbb{E}[X_t^2] < \infty$, then no measurable choice can be constructed, except $X = 0$ $\mathbb{P} \otimes dt$-a.s. Indeed, it would imply $\mathbb{E}\left[(\int_0^t X_s ds)^2\right] = \int_0^t \int_0^t \mathbb{E}[X_s X_u] ds\, du = 0$, hence X would be null.

We say that a deterministic function f is non-decreasing if $f(t) \geq f(s)$ for $t \geq s$. We say that a random variable Y is non-negative (resp. positive) if $Y \geq 0$ (resp. $Y > 0$) a.s. The equality between two random variables is always an a.s. equality. A process X is non-negative (resp. positive) if X_t is non-negative (resp. positive) for all t.

We shall often use independence of σ-fields. We say that σ-fields $\mathscr{G}$ and $\mathscr{F}$ are independent under $\mathbb{P}$, denoted $\mathscr{G} \perp\!\!\!\perp \mathscr{F}$, if for any $F \in \mathscr{F}$ and $G \in \mathscr{G}$ it holds that $\mathbb{P}(F \cap G) = \mathbb{P}(F)\mathbb{P}(G)$. More generally, for $\mathscr{A} \subset \mathscr{F}$, we say that σ-fields $\mathscr{G}$ and $\mathscr{F}$ are conditionally independent given $\mathscr{A}$ under $\mathbb{P}$, denoted $\mathscr{G} \perp\!\!\!\perp_{\mathscr{A}} \mathscr{F}$, if for any $F \in \mathscr{F}$ and $G \in \mathscr{G}$ it holds that $\mathbb{P}(F \cap G|\mathscr{A}) = \mathbb{P}(F|\mathscr{A})\mathbb{P}(G|\mathscr{A})$.

1.1.1 Path Properties

Definition 1.3 (a) A process X is **continuous** if the map $t \to X_t(\omega)$ is $\mathbb{P}$-a.s. continuous. A process X is **càdlàg**[1] if the map $t \to X_t(\omega)$ is right-continuous with left-limits $\mathbb{P}$-a.s.
(b) A process X is **increasing** if X is càdlàg, and the map $t \to X_t(\omega)$ is non-decreasing $\mathbb{P}$-a.s.
(c) A process X has **finite variation** if it is the difference of two non-negative increasing processes X^+ and X^-. A finite variation process X is **integrable** if $\mathbb{E}[X_\infty^+ + X_\infty^-] < \infty$, where $X_\infty^\pm = \lim_{t\to\infty} X_t^\pm$.

A process Y is a **modification** of a process X if, for any t, $\mathbb{P}(X_t = Y_t) = 1$. However, one needs a stronger assumption to be able to compare functionals of the processes. The process Y is **indistinguishable** from X if $\{\omega : X_t(\omega) = Y_t(\omega), \forall t\}$ is a measurable set and $\mathbb{P}(X_t = Y_t, \forall t) = 1$. If X and Y are modifications of each other and are càdlàg, they are indistinguishable.

For a càdlàg process, we write $X_{t-} := \lim_{s\to t, s<t} X_s$ for $t > 0$, $X_{0-} = 0$ and $\Delta X_t = X_t - X_{t-}$ for $t \geq 0$. For a σ-field $\mathscr{G}$, $X \in L^p(\mathscr{G})$ means that X is a $\mathscr{G}$-measurable random variable such that $\mathbb{E}[|X|^p] < \infty$.

---

[1] In French, right-continuous is <u>c</u>ontinu <u>à</u> <u>d</u>roite, and with left-limits is admettant des <u>l</u>imites <u>à</u> <u>g</u>auche. We shall also use càd for right-continuous and càglàd for left-continuous with right-limits. The use of these acronyms comes from P.-A. Meyer.

1.1.2 Random Times and Stopping Times

Definition 1.4 A **random time** τ is a random variable valued in $\overline{\mathbb{R}}^+ := [0, \infty]$.
An $\mathbb{F}$-**stopping time** is a random time such that for any $t \geq 0$, $\{\tau \leq t\} \in \mathscr{F}_t$.

If a random time τ satisfies $\{\tau < t\} \in \mathscr{F}_t$ for any $t \geq 0$, it is an $\mathbb{F}$-stopping time, since $\{\tau \leq t\} = \bigcap_{s>t}\{\tau < s\}$. Note that the right-continuity of the filtration $\mathbb{F}$ is essential here. If needed, we shall indicate which filtration is used, by writing that τ is an $\mathbb{F}$-stopping time.

The **graph** of a random time is given by

$$[\![\tau]\!] := \{(\omega, t) \in \Omega \times \mathbb{R}^+ : \tau(\omega) = t\}.$$

A stopping time τ is **predictable** if there exists an increasing sequence $(\tau_n, n \geq 0)$ of stopping times such that almost surely

(a) $\lim_n \tau_n = \tau$,

(b) $\tau_n < \tau$ for every n on the set $\{\tau > 0\}$.

The sequence (τ_n) is called an **announcing sequence.**

A stopping time τ is **accessible** if there exists a sequence $(T_n, n \geq 0)$ of predictable stopping times such that $[\![\tau]\!] \subset \cup_n [\![T_n]\!]$.

A stopping time τ is **totally inaccessible** if $\mathbb{P}(\tau = \vartheta < \infty) = 0$ for any predictable stopping time ϑ, or, equivalently, if for any increasing sequence of stopping times $(\tau_n, n \geq 0)$, one has $\mathbb{P}(\{\lim \tau_n = \tau\} \cap \{\cap_n \{\tau_n < \tau\}\}) = 0$.

Definition 1.5 If τ is an $\mathbb{F}$-stopping time, the σ-field $\mathscr{F}_\tau$ of events prior to τ, and the σ-field $\mathscr{F}_{\tau-}$ of events strictly prior to τ are defined as:

$$\mathscr{F}_\tau := \{F \in \mathscr{F}_\infty : F \cap \{\tau \leq t\} \in \mathscr{F}_t, \forall t\}$$

whereas $\mathscr{F}_{\tau-}$ is the smallest σ-field which contains $\mathscr{F}_0$ and all the sets of the form $F \cap \{t < \tau\}$ for $t > 0$ and $F \in \mathscr{F}_t$. Here $\mathscr{F}_{0-} = \mathscr{F}_0$.

The **restriction** of a random time to a given set $F \in \mathscr{F}$ is defined as

$$\tau_F(\omega) := \begin{cases} \tau(\omega) & \omega \in F \\ \infty & \omega \notin F. \end{cases} \tag{1.1}$$

If τ is an $\mathbb{F}$-stopping time and $F \in \mathscr{F}_\tau$, then τ_F is an $\mathbb{F}$-stopping time.

1.1.3 Predictable and Optional σ-Fields

For two random times τ and ϑ, the **stochastic interval** $]\!]\vartheta, \tau]\!]$ is the set

$$]\!]\vartheta, \tau]\!] := \{(\omega, t) \in \Omega \times \mathbb{R}^+ : \vartheta(\omega) < t \leq \tau(\omega)\}.$$

In an analogous way, we define the stochastic intervals $[\![\vartheta, \tau]\!]$, $[\![\vartheta, \tau[\![$ and $]\!]\vartheta, \tau[\![$.

Proposition 1.6 *(a) The* **optional** *σ-field $\mathscr{O}$ is the σ-field on $\Omega \times \mathbb{R}^+$ generated by $\mathbb{F}$-adapted càdlàg processes. The optional σ-field $\mathscr{O}$ is equal to the σ-field generated by all stochastic intervals $[\![\tau, \infty[\![$ where τ is an $\mathbb{F}$-stopping time.*
(b) The **predictable** *σ-field $\mathscr{P}$ is the σ-field on $\Omega \times \mathbb{R}^+$ generated by the $\mathbb{F}$-adapted càg (or continuous) processes. The predictable σ-field $\mathscr{P}$ is equal to the σ-field generated by all stochastic intervals $]\!]\tau, \infty[\![$, where τ is an $\mathbb{F}$-stopping time, and all sets $F \times \{0\}$ where $F \in \mathscr{F}_0$.*

If necessary, we will write $\mathscr{P}(\mathbb{F})$ for the predictable σ-field (or $\mathscr{O}(\mathbb{F})$ for the optional case), to emphasize the role of $\mathbb{F}$.

If all $\mathbb{F}$-martingales are continuous, then any $\mathbb{F}$-stopping time is predictable and the two σ-fields $\mathscr{P}(\mathbb{F})$ and $\mathscr{O}(\mathbb{F})$ are equal. This is the case if $\mathbb{F}$ is a Brownian filtration. In general

$$\mathscr{O}(\mathbb{F}) = \mathscr{P}(\mathbb{F}) \vee \sigma(\Delta N : N \text{ is a càdlàg } \mathbb{F}\text{-martingale}) .$$

A process X is said to be $\mathbb{F}$-**predictable** (resp. $\mathbb{F}$-**optional**) if it is $\mathscr{P}(\mathbb{F})$-measurable (resp. $\mathscr{O}(\mathbb{F})$-measurable).

If X is a predictable (resp. optional) process and τ is a stopping time, then the **stopped** process $X^\tau := (X^\tau_t = X_{t\wedge\tau}, t \geq 0)$ is also predictable (resp. optional). If X is a càdlàg $\mathbb{F}$-adapted process, then $(X_{t-}, t \geq 0)$ is a predictable process. A stopping time τ is predictable if and only if the (càdlàg) process $\mathbb{1}_{[\![0,\tau[\![}$ is predictable. If X is optional, then $\mathbb{E}\left[\int_0^\infty X_t dt\right] = \int_0^\infty \mathbb{E}[X_t]\, dt$, if all quantities are well-defined.

For a set $A \subset \Omega \times \mathbb{R}^+$, we denote by $\mathtt{i}(A)$ its projection on Ω, i.e., $\mathtt{i}(A) := \{\omega : \exists t \text{ s.t. } (\omega, t) \in A\}$. A set $A \subset \Omega \times \mathbb{R}^+$ is called evanescent if its projection $\mathtt{i}(A)$ is a $\mathbb{P}$-null set.

In the next theorem the very useful and important **section theorem** is recalled (see e.g. [112, Theorems 4.7 and 4.8]). We will refer to the first assertion as the optional section theorem and to the second assertion as the predictable section theorem.

Theorem 1.7 *(a) Let A be an $\mathbb{F}$-optional set. Then, for each $\varepsilon > 0$, there exists an $\mathbb{F}$-stopping time such that $[\![T]\!] \subset A$ and $\mathbb{P}(T < \infty) \geq \mathbb{P}(\mathtt{i}(A)) - \varepsilon$.*
(b) Let A be an $\mathbb{F}$-predictable set. Then, for each $\varepsilon > 0$, there exists an $\mathbb{F}$-predictable stopping time such that $[\![T]\!] \subset A$ and $\mathbb{P}(T < \infty) \geq \mathbb{P}(\mathtt{i}(A)) - \varepsilon$.

The section theorem has the following consequence. Let X and Y be two optional (resp. predictable) processes. If for every finite stopping time (resp. predictable stopping time) ϑ, one has $X_\vartheta = Y_\vartheta$ a.s., then the processes X and Y are indistinguishable.

For a random time τ, the σ-fields $\mathscr{F}_\tau$ and $\mathscr{F}_{\tau-}$ are defined as

$$\mathscr{F}_\tau := \sigma\{X_\tau \mathbb{1}_{\{\tau<\infty\}},\ X \text{ is an } \mathbb{F}\text{-}optional \text{ process}\}$$
$$\mathscr{F}_{\tau-} := \sigma\{X_\tau \mathbb{1}_{\{\tau<\infty\}},\ X \text{ is an } \mathbb{F}\text{-}predictable \text{ process}\}.$$

Obviously, $\mathscr{F}_{\tau-} \subset \mathscr{F}_\tau$ and $\tau \in \mathscr{F}_{\tau-}$. For stopping times, this definition coincides with Definition 1.5.

1.1.4 Martingales and Local Martingales

Definition 1.8 An $\mathbb{F}$-adapted process X is an $\mathbb{F}$-**martingale** if
(a) For any t, the r.v. X_t is integrable, i.e., $\mathbb{E}[|X_t|] < \infty$,
(b) for any $s \leq t$, one has $\mathbb{E}[X_t|\mathscr{F}_s] = X_s$.
An $\mathbb{F}$-adapted process X is an $\mathbb{F}$-**supermartingale** (resp. submartingale) if
(a) for any t, the r.v. X_t is integrable,
(b) for any $s \leq t$, one has $\mathbb{E}[X_t|\mathscr{F}_s] \leq X_s$ (resp. $\mathbb{E}[X_t|\mathscr{F}_s] \geq X_s$).
A martingale X is said to be **square integrable** if $\sup_t \mathbb{E}[X_t^2] < \infty$.

Let $\mathscr{D}$ be a class of processes. The *localized class* of $\mathscr{D}$, denoted $\mathscr{D}_{loc}$, is defined in the following way. A process X belongs to $\mathscr{D}_{loc}$ if X_0 is $\mathscr{F}_0$-measurable and there exists a sequence of $\mathbb{F}$-stopping times $(\tau_n, n \geq 0)$ such that:
(a) the sequence (τ_n) is increasing and $\lim_n \tau_n = \infty$ a.s.,
(b) for every n, the stopped process $X^{\tau_n} - X_0 := (X_{t\wedge\tau_n} - X_0, t \geq 0)$ belongs to $\mathscr{D}$.
A sequence of stopping times such that the two previous conditions hold is called a **localizing sequence**.

In particular we use the above localization to define locally square integrable martingales, locally bounded martingales, local supermartingales and locally integrable variation processes.

If X is a local martingale, it is always possible to choose the localizing sequence (τ_n) such that each martingale $X^{\tau_n} - X_0$ is uniformly integrable.

We denote by $\mathscr{M}_{loc}(\mathbb{F}, \mathbb{P})$ the space of $(\mathbb{F}, \mathbb{P})$-local martingales. If a local martingale is not a martingale, we say that it is a strict local martingale.

The following two results will be useful in our setting. Theorems 1.9 and 1.10 are standard results for continuous time (super)martingales and can be found, e.g., in [112, Theorems 2.46 and 2.47].

Theorem 1.9 *Let $\widetilde{\mathbb{F}}$ be a filtration, not necessary right-continuous, and $\mathbb{F}$ its right-continuous regularization. Then any càdlàg $\widetilde{\mathbb{F}}$-supermartingale X is also an $\mathbb{F}$-supermartingale.*

Proof Fix $s < t$ and take a decreasing sequence $(s_n)_{n\in\mathbb{N}}$ such that $s_n \searrow s$ and $s < s_n < t$. From our assumption $\mathscr{F}_s = \bigcap_{u>s} \widetilde{\mathscr{F}}_u$. Then, for any $A \in \mathscr{F}_s \subset \widetilde{\mathscr{F}}_{s_n}$ we have $\mathbb{E}[X_{s_n} 1\!\!1_A] \geq \mathbb{E}[X_t 1\!\!1_A]$. Since $\mathbb{E}[|X_s|] < \infty$, [112, Theorem 2.22] implies that the family $(X_{s_n})_n$ is uniformly integrable. Then, by the Lebesgue–Vitali convergence theorem [43, Theorem 4.5.4] and the fact that X has càd paths, we obtain that $\lim_{n\to\infty} \mathbb{E}[X_{s_n} 1\!\!1_A] = \mathbb{E}\left[\lim_{n\to\infty} X_{s_n} 1\!\!1_A\right] = \mathbb{E}[X_s 1\!\!1_A]$. Therefore $\mathbb{E}[X_s 1\!\!1_A] \geq \mathbb{E}[X_t 1\!\!1_A]$ and X is an $\mathbb{F}$-supermartingale. □

Theorem 1.10 *Let $\mathbb{F}$ be a right-continuous filtration. Then, an $\mathbb{F}$-supermartingale X has a càdlàg modification if and only if the mapping $t \mapsto \mathbb{E}[X_t]$ is right-continuous.*

The statement remains valid for $\mathbb{F}$-submartingales. In particular each $\mathbb{F}$-martingale has a càdlàg modification since the mapping $t \mapsto \mathbb{E}[X_t]$ is constant.

The following result on (super)martingale convergence is known as Lévy's theorem.

Theorem 1.11 *Let $\mathbb{F}$ be a filtration, not necessary right-continuous. Suppose that ξ is an integrable r.v. and X is an $\mathbb{F}$-supermartingale. Then $\mathbb{E}[\xi|\mathcal{F}_t] \to \mathbb{E}[\xi|\mathcal{F}_\infty]$ a.s. and in L^1 as $t \to \infty$; and, for $t_n \searrow t$, $X_{t_n} \to X_{t+}$ a.s. and in L^1 as $n \to \infty$.*

In particular one can apply the above theorem to a martingale $X_t = \mathbb{E}[\xi|\mathcal{F}_t]$ which ensures that, for $t_n \searrow t$, $\mathbb{E}[\xi|\mathcal{F}_{t_n}] \to \mathbb{E}[\xi|\mathcal{F}_{t+}]$ a.s and in L^1 as $n \to \infty$.

The right-continuity of a filtration, needed for the existence of càdlàg modification of martingales, is a delicate issue. It is of particular importance in the theory of enlargement of filtrations: it is not true in general that if $\mathbb{F}$ and $\widetilde{\mathbb{F}}$ are right-continuous, the filtration $\mathbb{F} \vee \widetilde{\mathbb{F}}$ is right-continuous as well. In particular, if ζ is a random variable and $\mathbb{F}$ a right-continuous filtration, $\mathbb{F} \vee \sigma(\zeta)$ may fail to be so. See bibliographic notes for references. However $\mathbb{F} \vee \widetilde{\mathbb{F}}$ turns to be right-continuous in the following special case.

Proposition 1.12 *Let $\mathbb{F}$ and $\widetilde{\mathbb{F}}$ be two right-continuous filtrations such that $\mathcal{F}_\infty$ and $\widetilde{\mathcal{F}}_\infty$ are independent. Denote $\mathbb{G} := \mathbb{F} \vee \widetilde{\mathbb{F}}$. Then:*
(a) for two bounded resp. $\mathcal{F}_\infty$ and $\widetilde{\mathcal{F}}_\infty$-measurable r.v.'s ζ and $\widetilde{\zeta}$ and $t \geq 0$,

$$\mathbb{E}[\zeta\widetilde{\zeta}|\mathcal{G}_t] = \mathbb{E}[\zeta|\mathcal{F}_t]\mathbb{E}[\widetilde{\zeta}|\widetilde{\mathcal{F}}_t];$$

(b) the filtration $\mathbb{G}$ is right-continuous.

Proof The assertion (a) indeed holds since for any $F \in \mathcal{F}_t$ and $\widetilde{F} \in \widetilde{\mathcal{F}}_t$, the independence hypothesis implies that

$$\begin{aligned} \mathbb{E}\left[\zeta\widetilde{\zeta}\mathbb{1}_{F\cap\widetilde{F}}\right] &= \mathbb{E}[\zeta\mathbb{1}_F]\,\mathbb{E}\left[\widetilde{\zeta}\mathbb{1}_{\widetilde{F}}\right] = \mathbb{E}\left[\mathbb{E}[\zeta|\mathcal{F}_t]\mathbb{1}_F\right]\,\mathbb{E}\left[\mathbb{E}[\widetilde{\zeta}|\widetilde{\mathcal{F}}_t]\mathbb{1}_{\widetilde{F}}\right] \\ &= \mathbb{E}\left[\mathbb{E}[\zeta|\mathcal{F}_t]\,\mathbb{E}[\widetilde{\zeta}|\widetilde{\mathcal{F}}_t]\mathbb{1}_{F\cap\widetilde{F}}\right]. \end{aligned}$$

(b) Note that for any two bounded resp. $\mathcal{F}_\infty$ and $\widetilde{\mathcal{F}}_\infty$-measurable r.v.'s ζ and $\widetilde{\zeta}$ and $t \geq 0$ it holds that

$$\mathbb{E}[\zeta\widetilde{\zeta}|\mathcal{G}_{t+}] = \lim_{t_n \searrow t} \mathbb{E}[\zeta|\mathcal{F}_{t_n}]\,\mathbb{E}[\widetilde{\zeta}|\widetilde{\mathcal{F}}_{t_n}] = \mathbb{E}[\zeta|\mathcal{F}_t]\,\mathbb{E}[\widetilde{\zeta}|\widetilde{\mathcal{F}}_t] = \mathbb{E}[\zeta\widetilde{\zeta}|\mathcal{G}_t]$$

where the first equality follows by Theorem 1.11, the second by the right-continuity of $\mathbb{F}$ and $\widetilde{\mathbb{F}}$ and the last by the assertion (a) Since r.v's of the above form $\zeta\widetilde{\zeta}$ generate all bounded $\mathcal{G}_\infty$-measurable r.v.'s, we conclude that for any bounded $\mathcal{G}_\infty$-measurable r.v. ξ it holds that $\mathbb{E}[\xi|\mathcal{G}_{t+}] = \mathbb{E}[\xi|\mathcal{G}_t]$. Hence $\mathbb{G}$ is right-continuous. □

1.2 Semimartingales

An $\mathbb{F}$-adapted càdlàg process X is called an $\mathbb{F}$-semimartingale if $X = X_0 + N + V$ where N is an $\mathbb{F}$-local martingale with $N_0 = 0$ and V is an $\mathbb{F}$-adapted process with finite variation and $V_0 = 0$. A decomposition where the process V is predictable is unique (if it exists) and called the **canonical decomposition** of X. In that case, X is called a **special semimartingale**. If X is a continuous semimartingale, then V (hence N) is continuous and X is a special semimartingale.

Definition 1.13 The $\mathbb{F}$-predictable finite variation part of a special semimartingale X is called **the $\mathbb{F}$-compensator** of X. It compensates X in the sense that $X - V$ is an $\mathbb{F}$-local martingale.

1.2.1 *The Bichteler–Dellacherie–Mokobodzki Theorem*

A simple integrand is a stochastic process of the form $H_t = \sum_{i=0}^{n} H_i \mathbb{1}_{]\!]T_i,T_{i+1}]\!]}(t)$ where $0 = T_0 \leq T_1 \leq \cdots \leq T_{n+1}$ is a finite sequence of stopping times and H_i is an $\mathcal{F}_{T_i}$-measurable random variable for each $0 \leq i \leq n$. Given an adapted càdlàg process X and a simple integrand H, we define the integral process $I_X(H)$ as

$$I_X(H)_t := \sum_{i=0}^{n} H_i (X_{T_{i+1}\wedge t} - X_{T_i \wedge t}).$$

A process X is called a **good integrator** if for any sequence of simple integrands H^n such that, for any $t \geq 0$, $\|H_t^n\| := \sup_{\omega, s\leq t} |H_s^n(\omega)| \to 0$ when n goes to infinity, we have that $I_X(H^n)$ converges in probability on compacts to 0, i.e., for each $t \geq 0$ and each $\varepsilon > 0$ one has $\lim_{n\to\infty} \mathbb{P}(|I_X(H^n)_t| > \varepsilon) = 0$.

Theorem 1.14 (Bichteler–Dellacherie–Mokobodzki Theorem) *A càdlàg adapted process X is a semimartingale if and only if X is a good integrator.*

1.2.2 *Integration by Parts and Covariation Processes*

For processes H and X we write $\int_0^t H_s dX_s := \int_{(0,t]} H_s dX_s$ whenever the integral exists as a pathwise integral of stochastic integral. For a (bounded) predictable process H and a semimartingale X with decomposition $N + V$, we denote by $H \cdot X_t$ the stochastic integral

$$H \cdot X_t := \int_0^t H_s dX_s = \int_0^t H_s dN_s + \int_0^t H_s dV_s .$$

The process $H \cdot X$ is a semimartingale with decomposition $H \cdot N + H \cdot V$.

The martingale part of X admits a unique decomposition as $N = N^c + N^d$ where N^c is continuous, $N_0^c = 0$ and N^d is a purely discontinuous martingale, $N_0^d = 0$. The process N^c is denoted in the literature as X^c (even if this notation can be misleading!). We recall that two local martingales are **orthogonal** if their product is a local martingale and that a **purely discontinuous martingale** is a martingale orthogonal to all continuous martingales.

Definition 1.15 Let X and Y be càdlàg local martingales and define $X_{0-} = Y_{0-} = 0$.
(a) The **quadratic covariation process** $[X, Y]$ is the unique càdlàg finite variation process such that

- $XY - [X, Y]$ is a local martingale,
- $\Delta[X, Y] = \Delta X\, \Delta Y$, (define $[X, Y]_{0-} = 0$)

We shall denote $[X] := [X, X]$ and call it the **quadratic variation process** of X.
(b) The **predictable quadratic covariation process** of two càdlàg local martingales X and Y is (if it exists) the predictable finite variation process $\langle X, Y\rangle$ such that $XY - \langle X, Y\rangle$ is a local martingale. Under the condition $\langle X, Y\rangle_0 = X_0 Y_0$, it is unique. We shall denote $\langle X\rangle := \langle X, X\rangle$ and call it **predictable quadratic variation process** of X (if it exists).

Note that the above definition in particular implies that $[X, Y]_0 = X_0 Y_0$. If X and Y are continuous martingales, then $\langle X, Y\rangle$ exists and $\langle X, Y\rangle = [X, Y]$. Two càdlàg local martingales X and Y are orthogonal if and only if $[X, Y]$ is a local martingale. The processes $[X]$ and $\langle X\rangle$ are increasing.

If X and Y are semimartingales, their quadratic covariation process is defined as

$$[X, Y]_t := X_0 Y_0 + \langle X^c, Y^c\rangle_t + \sum_{0<s\le t} \Delta X_s\, \Delta Y_s = \langle X^c, Y^c\rangle_t + \sum_{0\le s\le t} \Delta X_s\, \Delta Y_s,$$

where X^c, introduced before Definition 1.15, is a continuous local martingale s.t. $X_0^c = 0$. The predictable covariation process of two semimartingales is defined (if it exists) as the predictable finite variation process $\langle X, Y\rangle$ such that $[X, Y] - \langle X, Y\rangle$ is a local martingale and $\langle X, Y\rangle_0 = X_0 Y_0$. We refer to Sect. 1.3.2 for more details.

If X and Y are semimartingales with respect to $\mathbb{F}$ and $\mathbb{G}$, then $[X, Y]$ is the same in both filtrations, but $\langle X, Y\rangle^{\mathbb{F}}$ and $\langle X, Y\rangle^{\mathbb{G}}$ may differ (and often do).

Proposition 1.16 *Let X, Y be two semimartingales and V a finite variation process.*
(a) The integration by parts formula for semimartingales is

$$X_t Y_t = \int_0^t X_{s-} dY_s + \int_0^t Y_{s-} dX_s + [X, Y]_t\,. \tag{1.2}$$

(b) The quadratic covariation process of X and V equals $[X, V]_t = X_0 V_0 + \Delta X \cdot V_t = \int_{[0,t]} \Delta X_s dV_s$ and

$$X_t V_t = \int_0^t X_s dV_s + \int_0^t V_{s-} dX_s. \tag{1.3}$$

(c) If additionally V is a predictable process, then $[X, V] = X_0 V_0 + \Delta V \cdot X$ *and*

$$X_t V_t = \int_0^t X_{s-} dV_s + \int_0^t V_s dX_s. \tag{1.4}$$

Equalities (1.3) and (1.4) are known under the name **Yoeurp's Lemma** [212].

1.2.3 Martingale and Semimartingale Spaces

Definition 1.17 For any $p \in [1, \infty]$, we denote[2] by $H^p(\mathbb{F})$ the Banach space consisting of $\mathbb{F}$-local martingales X equipped with the norm $\| \cdot \|_{H^p(\mathbb{F})}$ given by:

$$\|X\|_{H^p(\mathbb{F})} := \|[X]_\infty^{1/2}\|_{L^p} < \infty .$$

Recall the Burkholder–Davis–Gundy inequality stating that, for each $p \in [1, \infty)$ there exist two constants c_p and C_p such that for any local martingale X and stopping time T it holds that

$$c_p \mathbb{E}\left[[X]_T^{p/2}\right] \leq \mathbb{E}\left[\sup_{t \leq T} |X_t|^p\right] \leq C_p \mathbb{E}\left[[X]_T^{p/2}\right].$$

The Burkholder–Davis–Gundy inequality provides an equivalent norm to $\| \cdot \|_{H^p(\mathbb{F})}$.

Definition 1.18 Let X be a uniformly integrable $\mathbb{F}$-martingale and let $p \in [1, \infty)$. We say that X belongs to the space $BMO_p(\mathbb{F})$ if there exists a constant C such that

$$\mathbb{E}\left[|X_\infty - X_{T-} \mathbb{1}_{\{T>0\}}|^p | \mathcal{F}_T\right] \leq C^p \quad a.s. \quad \text{for any stopping time T.}$$

The $BMO_p(\mathbb{F})$-norm of X, denoted by $\|X\|_{BMO_p(\mathbb{F})}$, is the smallest constant C with this property.

Clearly, when p increases, norms BMO_p are stronger and stronger, so spaces BMO_p are smaller and smaller. But, as $BMO_1 \subset BMO_n$ for n integer ([79, VI.109 formulae (109.7)]), the spaces BMO_p are all the same with norms $\| \cdot \|_{BMO_p}$ all equivalent. We simply write BMO to denote this space and $\| \cdot \|_{BMO}$ for the norm in BMO_2, the most frequently used. The following sequence of inclusion holds

$$H^\infty \subset BMO \subset H^p \subset H^1 \subset \{\text{UI martingales}\} \subset \{\text{local martingales}\}.$$

[2] H comes from Hardy.

Every local martingale is locally in H^1. Bounded martingales are dense in H^1 in the norm $\|\cdot\|_{H^1}$.

The essential results are the duality (H^1, BMO) and Fefferman's inequality. Recall that BMO is the dual space of H^1, i.e., for any bounded linear mapping φ on H^1 there exists a BMO-martingale Y such that $\varphi(X) = \mathbb{E}[[X, Y]_\infty]$ for any $X \in H^1$. In particular, there exists a constant C such that for any local martingales X and Y,

$$\mathbb{E}\left[\int_{[0,\infty)} |d[X, Y]_s|\right] \leq C\|X\|_{H^1(\mathbb{F})}\|Y\|_{BMO(\mathbb{F})}. \tag{1.5}$$

The Fefferman inequality in particular implies that, if Y is a BMO-martingale and X is a local martingale, then the quadratic covariation process $[X, Y]$ has locally integrable variation, since X is locally in H^1. We shall see in Sect. 1.3.2 that, as a consequence, if Y is a BMO-martingale and X is a local martingale, then the predictable quadratic covariation process $\langle X, Y\rangle^{\mathbb{F}}$ exists.

1.2.4 Hypotheses $\mathscr{H}$ and $\mathscr{H}'$

In general, if $\mathbb{G} = (\mathscr{G}_t, t \geq 0)$ is a filtration larger than $\mathbb{F} = (\mathscr{F}_t, t \geq 0)$, i.e., for each $t \geq 0$, $\mathscr{F}_t \subset \mathscr{G}_t$ (we write $\mathbb{F} \subset \mathbb{G}$), it is not always true that an $\mathbb{F}$-martingale is a $\mathbb{G}$-martingale. It is not even true that an $\mathbb{F}$-martingale is always a $\mathbb{G}$-semimartingale. One of the goals of this book is to give conditions so that these properties hold.

Example 1.19 (a) Let $\mathscr{G}_t = \mathscr{F}_\infty$, $\forall t \geq 0$. Then, the only $\mathbb{F}$-martingales which are $\mathbb{G}$-martingales are constant processes.
(b) An interesting example is the Azéma martingale ζ, defined as follows (see also [25]). Let B be a Brownian motion and $g_t = \sup\{s \leq t, B_s = 0\}$. The process

$$\zeta_t = \text{sgn}\,(B_t)\sqrt{t - g_t}, \quad t \geq 0$$

is a martingale in its natural filtration $\mathbb{F}^\zeta$, and $\mathbb{F}^\zeta \subset \mathbb{F}^B$. This càdlàg $\mathbb{F}^\zeta$-martingale is not an $\mathbb{F}^B$-martingale since it is purely discontinuous and $\mathbb{F}^B$-martingales are continuous. It is not even an $\mathbb{F}^B$-semimartingale as, additionally, it has infinite variation.
(c) A Brownian motion B is not a semimartingale in the larger filtration $\mathbb{G}$, defined as $\mathscr{G}_t = \mathscr{F}^B_{t+\delta}$ with $\delta > 0$. Indeed, the Brownian motion is $\mathscr{G}_0$-measurable on the interval $[0, \delta]$ but it has infinite variation. Then, B is not a $\widehat{\mathbb{G}}$-semimartingale, where $\widehat{\mathscr{G}}_t = \mathscr{F}^B_t \vee \mathscr{F}^B_\delta$, as $\widehat{\mathscr{G}}_0 = \mathscr{G}_0$ and all $\widehat{\mathbb{G}}$-martingales are constant on $[0, \delta]$. Since $\widehat{\mathbb{G}} \subset \mathbb{G}$, the Bichteler–Dellacherie–Mokobodzki Theorem 1.14 implies that B is not a $\mathbb{G}$-semimartingale (see also Theorem 1.25). See [92, 188] for details.
(d) Starting with a random time τ, and the smallest right-continuous filtration $\mathbb{A}$ making τ a stopping time, Jeulin [139] shows (with an explicit construction) that there exists a process which is a continuous martingale w.r.t. its natural filtration and which is $\mathbb{A}$-adapted, if and only if the law of τ has a diffuse part. However, the

natural filtration of this process is strictly smaller than $\mathbb{A}$ and this process is not an $\mathbb{A}$-semimartingale.

Definition 1.20 Let $\mathbb{F} \subset \mathbb{G}$. Then:
(a) The hypothesis $\mathscr{H}$ (or immersion) holds between $\mathbb{F}$ and $\mathbb{G}$, or $\mathbb{F}$ is immersed in $\mathbb{G}$, if any càdlàg $\mathbb{F}$-martingale is a $\mathbb{G}$-martingale. We denote it by $\mathbb{F} \hookrightarrow \mathbb{G}$.
(b) The hypothesis $\mathscr{H}'$ holds between $\mathbb{F}$ and $\mathbb{G}$ if any càdlàg $\mathbb{F}$-martingale is a $\mathbb{G}$-semimartingale.

We start with simple situations where the immersion holds.

Proposition 1.21 *Let $\mathbb{F}$ and $\widetilde{\mathbb{F}}$ be two filtrations, such that $\mathscr{F}_\infty$ and $\widetilde{\mathscr{F}}_\infty$ are independent. Then, denoting $\mathbb{G} := \mathbb{F} \vee \widetilde{\mathbb{F}}$, $\mathbb{F} \hookrightarrow \mathbb{G}$ and $\widetilde{\mathbb{F}} \hookrightarrow \mathbb{G}$.*

Proof By localization, it is enough to show that any u.i. $\mathbb{F}$-martingale X is a $\mathbb{G}$-martingale which easily follows by independence, i.e., $\mathbb{E}[X_\infty|\mathscr{G}_t] = \mathbb{E}[X_\infty|\mathscr{F}_t]$. □

Proposition 1.22 *Let $\mathbb{F}$ be a filtration and ζ be a r.v. Then, $\mathbb{F} \hookrightarrow \mathbb{F} \vee \sigma(\zeta)$ if and only if ζ and $\mathscr{F}_\infty$ are conditionally independent given $\mathscr{F}_0$ (denoted $\sigma(\zeta) \perp\!\!\!\perp_{\mathscr{F}_0} \mathscr{F}_\infty$). In particular, $\mathbb{F} \hookrightarrow \mathbb{F} \vee \sigma(\zeta)$ if ζ and $\mathscr{F}_\infty$ are independent (denoted $\sigma(\zeta) \perp\!\!\!\perp \mathscr{F}_\infty$).*

Proof Let f be a bounded Borel function and ξ be a bounded $\mathscr{F}_\infty$-measurable r.v. The necessary condition follows by

$$\begin{aligned}\mathbb{E}[f(\zeta)\xi|\mathscr{F}_0] &= \mathbb{E}[f(\zeta)\mathbb{E}[\xi|\mathscr{F}_0 \vee \sigma(\zeta)]|\mathscr{F}_0] = \mathbb{E}[f(\zeta)\mathbb{E}[\xi|\mathscr{F}_0]|\mathscr{F}_0] \\ &= \mathbb{E}[f(\zeta)|\mathscr{F}_0]\mathbb{E}[\xi|\mathscr{F}_0]\end{aligned}$$

where the second equality holds due to immersion. To show that it is a sufficient condition note that $\sigma(\zeta) \perp\!\!\!\perp_{\mathscr{F}_0} \mathscr{F}_\infty$ implies $\sigma(\zeta) \perp\!\!\!\perp_{\mathscr{F}_t} \mathscr{F}_\infty$ since $\mathscr{F}_0 \subset \mathscr{F}_t \subset \mathscr{F}_\infty$. Then, for a bounded Borel function f, a bounded $\mathscr{F}_\infty$-measurable r.v. ξ and $F \in \mathscr{F}_t$ conditional independence implies

$$\mathbb{E}[\xi f(\zeta)\mathbb{1}_F] = \mathbb{E}\left[\mathbb{E}[\xi|\mathscr{F}_t]\mathbb{E}[f(\zeta)|\mathscr{F}_t]\mathbb{1}_F\right] = \mathbb{E}\left[\mathbb{E}[\xi|\mathscr{F}_t]f(\zeta)\mathbb{1}_F\right].$$

Since r.v.'s of the form $f(\zeta)\mathbb{1}_F$ generate σ-field $\mathscr{F}_t \vee \sigma(\zeta)$ we conclude that $\mathbb{E}[\xi|\mathscr{F}_t \vee \sigma(\zeta)] = \mathbb{E}[\xi|\mathscr{F}_t]$ which implies immersion.

The last assertion comes from the fact that $\sigma(\zeta) \perp\!\!\!\perp \mathscr{F}_\infty$ implies $\sigma(\zeta) \perp\!\!\!\perp_{\mathscr{F}_0} \mathscr{F}_\infty$ since $\mathscr{F}_0 \subset \mathscr{F}_\infty$. □

Proposition 1.23 *Let $\mathbb{F}$ be a filtration and ζ be a r.v. such that ζ and $\mathscr{F}_\infty$ are conditionally independent given $\mathscr{F}_0$. Suppose that an $\mathbb{F}$-local martingale X has the PRP in $\mathbb{F}$, i.e., for any $\mathbb{F}$-local martingale Y there exists an $\mathbb{F}$-predictable process φ such that $Y = Y_0 + \varphi \cdot X$. Then, denoting $\mathbb{G} := \mathbb{F} \vee \sigma(\zeta)$, X has the PRP in $\mathbb{G}$.*

Proof Let us consider a $\mathbb{G}$-martingale Y of the form $Y_t = \mathbb{E}[\xi f(\zeta)|\mathscr{G}_t]$ where ξ is a bounded $\mathscr{F}_\infty$-measurable r.v. and f is bounded Borel function. Then, since by independence $Y_t = f(\zeta)\mathbb{E}[\xi|\mathscr{F}_t]$, there exists an $\mathbb{F}$-predictable process φ such that

$Y = Y_0 + f(\zeta)(\varphi \cdot X)$. Note that the process $f(\zeta)\varphi$ is $\mathbb{G}$-predictable and the $\mathbb{G}$-stochastic integral $(f(\zeta)\varphi) \cdot X$ exists since the $\mathbb{F}$-stochastic integral $\varphi \cdot X$ exists and f is bounded. Since the $\mathbb{G}$-martingales of the above form generate all bounded $\mathbb{G}$-martingales and bounded $\mathbb{G}$-martingales are dense in the space of u.i. $\mathbb{G}$-martingales, the results for an arbitrary $\mathbb{G}$-local martingale follows by localization. □

The following simple result exploits conditional expectation properties.

Proposition 1.24 *Let $\mathbb{F}$ and $\mathbb{G}$ be two filtrations such that $\mathbb{F} \subset \mathbb{G}$.*
(a) If X is a $\mathbb{G}$-martingale which is $\mathbb{F}$-adapted, then it is also an $\mathbb{F}$-martingale. In particular, if $\mathbb{F} \hookrightarrow \mathbb{G}$, and $\mathbb{F} \subset \widetilde{\mathbb{F}} \subset \mathbb{G}$, then, $\mathbb{F} \hookrightarrow \widetilde{\mathbb{F}}$.
(b) If Y is a uniformly integrable $\mathbb{F}$-martingale, then there exists a $\mathbb{G}$-martingale X, such that $\mathbb{E}[X_t|\mathscr{F}_t] = Y_t$, $t \geq 0$.

Proof (a) Let X be an $\mathbb{F}$-adapted $\mathbb{G}$-martingale. Then, for $s < t$, using the tower property,

$$\mathbb{E}[X_t|\mathscr{F}_s] = \mathbb{E}[\mathbb{E}[X_t|\mathscr{G}_s]|\mathscr{F}_s] = \mathbb{E}[X_s|\mathscr{F}_s] = X_s$$

we conclude that it is an $\mathbb{F}$-martingale. In particular, since $\mathbb{F} \hookrightarrow \mathbb{G}$, any $\mathbb{F}$-martingale is an $\widetilde{\mathbb{F}}$-adapted $\mathbb{G}$-martingale, thus it is an $\widetilde{\mathbb{F}}$-martingale.
(b) Since Y is uniformly integrable, Y_∞ exists and we introduce the $\mathbb{G}$-martingale X by $X_t := \mathbb{E}[Y_\infty|\mathscr{G}_t]$. Then, from the tower property $\mathbb{E}\,[X_t|\mathscr{F}_t] = \mathbb{E}[Y_\infty|\mathscr{F}_t] = Y_t$.□

We recall a fundamental result due to Stricker [203].

Theorem 1.25 *Let $\mathbb{F}$ and $\mathbb{G}$ be two filtrations such that $\mathbb{F} \subset \mathbb{G}$. A $\mathbb{G}$-semimartingale X which is $\mathbb{F}$-adapted, is also an $\mathbb{F}$-semimartingale.*

Proof The claim comes from Theorem 1.14, since $\mathbb{F}$-simple integrands are also $\mathbb{G}$-simple integrands. □

Proposition 1.26 *Let $\mathbb{F} \subset \mathbb{G}$ and X be an $\mathbb{F}$-local martingale which is also a $\mathbb{G}$-semimartingale.*
(a) Then X is a special $\mathbb{G}$-semimartingale.
(b) Denote $X = X^{\mathbb{G}} + V^{\mathbb{G}}$ its $\mathbb{G}$-canonical decomposition and let H be an $\mathbb{F}$-predictable process such that $\left(\int_0^\cdot H_s^2 d[X^{\mathbb{G}}]_s\right)^{1/2}$ is locally integrable. Then, $H \cdot X$ is a $\mathbb{G}$-semimartingale if and only if $\int_0^\cdot H_s dV_s^{\mathbb{G}}$ exists as a pathwise Stieltjes integral. In this case, the canonical decomposition of $H \cdot X$ is $H \cdot X^{\mathbb{G}} + H \cdot V^{\mathbb{G}}$.

Proof (a) Recall that, by [112, Theorem 8.6] a $\mathbb{G}$-semimartingale X is a special $\mathbb{G}$-semimartingale if and only if the process $\sqrt{[X]}$ is $\mathbb{G}$-locally integrable or if and only if the process $X^*_\cdot = \sup_{s\leq \cdot} X_s$ is $\mathbb{G}$-locally integrable. Since X is a special $\mathbb{F}$-semimartingale, $\sqrt{[X]}$ (and X^* as well) is $\mathbb{F}$-locally integrable, hence it is also $\mathbb{G}$-locally integrable as $\mathbb{F} \subset \mathbb{G}$. Thus X it is necessarily a special $\mathbb{G}$-semimartingale.
(b) See [138, Proposition 2.1]. □

From our perspective, it is important to investigate whether the stochastic integral is invariant w.r.t. the change of filtration. It is still an open problem for two non-nested incomparable filtrations and we refer the reader to Zheng [221] and Slud [195]. However, in an enlargement of filtration framework, the following result holds (see [138, Corollary (1,18)] for the proof).

Proposition 1.27 *Assume $\mathbb{F}$ and $\mathbb{G}$ are two filtrations such that $\mathbb{F} \subset \mathbb{G}$. Let X be an $\mathbb{F}$-semimartingale and a $\mathbb{G}$-semimartingale. For an $\mathbb{F}$-predictable bounded process H, stochastic integrals w.r.t. X computed respectively in $\mathbb{F}$ and $\mathbb{G}$ coincide.*

The following theorem concerns two *overlapping* filtrations and provides a useful procedure for checking if hypothesis $\mathcal{H}'$ holds.

Definition 1.28 Let $\mathbb{H}$ and $\mathbb{G}$ be two filtrations. Let T and S be two $\mathbb{H}$-stopping times which are also $\mathbb{G}$-stopping times. We say that $\mathbb{H}$ and $\mathbb{G}$ coincide on $[\![T, S[\![$ if

1. for each $t \geq 0$ and each $G \in \mathcal{G}_t$ there exists $H \in \mathcal{H}_t$ such that $\{T \leq t < S\} \cap G = \{T \leq t < S\} \cap H$
2. for each $t \geq 0$ and each $H \in \mathcal{H}_t$ there exists $G \in \mathcal{G}_t$ such that $\{T \leq t < S\} \cap H = \{T \leq t < S\} \cap G$.

Coincidence on other stochastic intervals, $[\![T, S]\!]$, $]\!]T, S]\!]$ and $]\!]T, S[\![$, is analogous.

Proposition 1.29 *Let $\mathbb{H}$ and $\mathbb{G}$ be two filtrations. Let T and S be two $\mathbb{H}$-stopping times and $\mathbb{G}$-stopping times. Assume that $\mathbb{H}$ and $\mathbb{G}$ coincide on $[\![T, S[\![$. Then, for a process X such that $X = \mathbb{1}_{]\!]T,S]\!]} \cdot X$, the following assertions hold.*
(a) The process X is a $\mathbb{H}$-(super, sub)martingale if and only if it is a $\mathbb{G}$-(super, sub)martingale.
(b) Let ϑ be a $\mathbb{H}$-stopping time. Then $(\vartheta \vee T) \wedge S$ is a $\mathbb{G}$-stopping time.
(c) The process X is a $\mathbb{H}$-local martingale if and only if it is a $\mathbb{G}$-local martingale.

Proof Coincidence of filtrations yields (a). Indeed, let X satisfy $X = \mathbb{1}_{]\!]T,S]\!]} \cdot X = X^S - X^T$, where X^S (resp. X^T) denotes the process X stopped at S (resp. T), and be $\mathbb{G}$-adapted (or equivalently $\mathbb{H}$-adapted), then, for $t \geq s$,

$$\mathbb{E}[X_t|\mathcal{G}_s] = \mathbb{E}[X_t \mathbb{1}_{\{s \geq S\}}|\mathcal{G}_s] + \mathbb{E}[X_t \mathbb{1}_{\{T \leq s < S\}}|\mathcal{G}_s] = \mathbb{E}[X_t|\mathcal{H}_s]$$

by coincidence and the fact that $X_t \mathbb{1}_{\{s \geq S\}} = X_S \mathbb{1}_{\{s \geq S\}}$ is $\mathcal{G}_s$- and $\mathcal{H}_s$-measurable. To prove assertion (b) we assume that ϑ is a $\mathbb{H}$-stopping time. Then, $(\vartheta \vee T) \wedge S$ is a $\mathbb{G}$-stopping time if and only if for each $t \geq 0$, $\{(\vartheta \vee T) > t, S > t\} \in \mathcal{G}_t$, equivalently $\{\vartheta \leq t, T \leq t, S > t\} \in \mathcal{G}_t$ which holds from coincidence of filtrations. Assertion (c) is shown as the combination of (a) and (b). □

1.2.5 Change of Probability Measure and Girsanov's Theorem

If Y is a semimartingale, the solution of $dL_t = L_{t-}dY_t$, $L_0 = 1$ is called the **Doléans-Dade exponential** of Y and is given by

$$\mathscr{E}(Y)_t := \exp\left(Y_t - Y_0 - \frac{1}{2}\langle Y^c\rangle_t\right) \prod_{s\leq t}(1+\Delta Y_s)e^{-\Delta Y_s}.$$

If Y is a local martingale, $\mathscr{E}(Y)$ is a local martingale. If additionally $L = \mathscr{E}(Y)$ is positive (this is the case if and only if $\Delta Y > -1$), it is a supermartingale; moreover, if $\mathbb{E}[L_t] = 1$, $\forall t$, it is a martingale. Furthermore, if L is a positive uniformly integrable martingale, one can define a probability measure $\mathbb{Q}$ on $\mathscr{F}_\infty$ as $d\mathbb{Q} = L_\infty d\mathbb{P}$.

Theorem 1.30 (Girsanov's Theorem) *Let $\mathbb{Q}$ be a probability measure equivalent to $\mathbb{P}$ on $\mathscr{F}_\infty$, with Radon–Nikodym density $\frac{d\mathbb{Q}}{d\mathbb{P}} = L_\infty$. Define an $(\mathbb{F},\mathbb{P})$-local martingale $L_t := \mathbb{E}_\mathbb{P}[L_\infty|\mathscr{F}_t]$. Then, for any $(\mathbb{F},\mathbb{P})$-local martingale X, the process $X^\mathbb{Q}$, given by $X^\mathbb{Q} := X - \frac{1}{L}\cdot[X,L]$, is an $(\mathbb{F},\mathbb{Q})$-local martingale.*

If additionally $\langle X,L\rangle^\mathbb{F}$ exists, the process $X - \frac{1}{L_-}\cdot\langle X,L\rangle^\mathbb{F}$ is an $(\mathbb{F},\mathbb{Q})$-local martingale.

Moreover, if Y is an $(\mathbb{F},\mathbb{Q})$-local martingale, then YL is an $(\mathbb{F},\mathbb{P})$-local martingale.

1.2.6 Decompositions of Supermartingales

An adapted process X is said to be of[3] class (D) if the collection $X_\tau \mathbb{1}_{\{\tau<\infty\}}$, where τ runs in the family of stopping times, is uniformly integrable. In particular, if X is bounded, X is of class (D).

In the following proposition **the additive Doob–Meyer decomposition** of a càdlàg supermartingale of class (D) is given.

Proposition 1.31 *If X is a càdlàg supermartingale of class (D), there exists a unique increasing, integrable and predictable process V such that $V_0 = 0$ and $X_t = \mathbb{E}[V_\infty - V_t|\mathscr{F}_t]$. In particular, any càdlàg supermartingale X of class (D) can be written as $X = \mu - V$ where μ is a uniformly integrable martingale. The decomposition is unique.*

In the following proposition **the predictable multiplicative decomposition** of a càdlàg positive supermartingale of class (D) is given.

[3] Class (D) is in honor of Doob.

Proposition 1.32 *Let X be a positive càdlàg supermartingale of class (D). There exists a unique pair of a local martingale N and a predictable decreasing process D such that $X = ND$ and $D_0 = X_0$.*

Proof Let $X = \mu - V$ be the Doob–Meyer decomposition of the supermartingale X. Since the process $\frac{1}{X_-}$ is locally bounded, the stochastic integral $\widehat{D} := \frac{1}{X_-} \cdot V$ exists and is a predictable, non-negative and increasing process. The process $\widehat{D}$ is also locally integrable and positive on $[\![T_0, \infty[\![$ where $T_0 := \inf\{t : V_t > 0\}$. Moreover $X_-(1 - \Delta\widehat{D}) = X_- - \Delta V = X - \Delta\mu$ and, for any predictable stopping time T, we have $X_{T-} - \Delta\widehat{D}_T = X_T - \Delta\mu_T = X_T > 0$ since μ is a martingale (see also Example 1.33). Then, by the predictable section theorem, we conclude that $1 - \Delta\widehat{D} > 0$. The process D given by $D := X_0 \mathscr{E}(-\widehat{D})$ is then positive, predictable, decreasing and satisfies $D_0 = X_0$. Let us show that the process $N := \frac{X}{D}$ is a local martingale. By Proposition 1.16 (c) and the definition of the Doléans-Dade exponential,

$$
\begin{aligned}
N &= 1 + \frac{1}{D} \cdot X + X_- \cdot \frac{1}{D} = 1 + \frac{1}{D} \cdot \mu - \frac{1}{D} \cdot V - \frac{X_-}{D_-^2} \cdot D - \sum \frac{\Delta D X_-}{D_-} \Delta\frac{1}{D} \\
&= 1 + \frac{1}{D} \cdot \mu - \frac{1}{D} \cdot V - \frac{1}{D_-} \cdot V + \sum \Delta V \Delta\frac{1}{D} = 1 + \frac{1}{D} \cdot \mu,
\end{aligned}
$$

where $\sum$ runs over all $s > 0$. This shows that N is indeed a local martingale and the proof is completed. □

1.3 Projections, Dual Projections and Applications

1.3.1 Projections

Let $\mathbb{F}$ be a given filtration and X a bounded (or non-negative) process not necessarily $\mathbb{F}$-adapted. The **$\mathbb{F}$-optional projection** of X is the unique $\mathbb{F}$-optional process ${}^oX$ which, for any $\mathbb{F}$-stopping time T, satisfies

$$\mathbb{E}[X_T \mathbb{1}_{\{T<\infty\}}] = \mathbb{E}[{}^oX_T \mathbb{1}_{\{T<\infty\}}]\,. \tag{1.6}$$

For any $\mathbb{F}$-stopping time T, let $\Gamma \in \mathscr{F}_T$ and apply the equality (1.6) to the stopping time $T_\Gamma = T\mathbb{1}_\Gamma + \infty\mathbb{1}_{\Gamma^c}$. We get the reinforced identity

$$\mathbb{E}[X_T \mathbb{1}_{\{T<\infty\}} | \mathscr{F}_T] = {}^oX_T \mathbb{1}_{\{T<\infty\}}\,. \tag{1.7}$$

In particular for any t, $\mathbb{E}[X_t|\mathscr{F}_t] = {}^oX_t$. For any bounded variation $\mathbb{F}$-optional process V, one has

$$\mathbb{E}\left[\int_{[0,\infty)} H_t dV_t\right] = \mathbb{E}\left[\int_{[0,\infty)} {}^o H_t dV_t\right]. \tag{1.8}$$

Likewise, the **$\mathbb{F}$-predictable projection** of the process X is the unique predictable process ${}^p X$ which, for any *$\mathbb{F}$-predictable stopping time* ϑ, satisfies

$$\mathbb{E}[X_\vartheta \mathbb{1}_{\{\vartheta<\infty\}}] = \mathbb{E}\left[{}^p X_\vartheta \mathbb{1}_{\{\vartheta<\infty\}}\right].$$

Like previously, we get the identity $\mathbb{E}[X_\vartheta \mathbb{1}_{\{\vartheta<\infty\}} | \mathscr{F}_{\vartheta-}] = {}^p X_\vartheta \mathbb{1}_{\{\vartheta<\infty\}}$, for any $\mathbb{F}$-predictable stopping time ϑ. For any bounded variation $\mathbb{F}$-predictable process V

$$\mathbb{E}\left[\int_{[0,\infty)} H_t dV_t\right] = \mathbb{E}\left[\int_{[0,\infty)} {}^p H_t dV_t\right]. \tag{1.9}$$

To avoid a confusion in the case where many filtrations are involved, we will use the notation ${}^{o,\mathbb{F}}X$ (resp. ${}^{p,\mathbb{F}}X$) for the $\mathbb{F}$-optional (resp. $\mathbb{F}$-predictable) projection.

Example 1.33 If X is a càdlàg uniformly integrable martingale, its predictable projection is X_-.

1.3.2 Dual Projections

Recall that for any finite variation process V, we adopt the convention that $V_{0-} = 0$ implying $\Delta V_0 = V_0$. This allows us to consider Stieltjes' integration w.r.t. V on the interval $[0, \infty)$. More precisely $\int_{[0,\infty)} H_s dV_s = H_0 V_0 + \int_0^\infty H_s dV_s$.

Proposition 1.34 *Let V be a locally integrable variation process not necessarily $\mathbb{F}$-adapted.*
(a) There exists a unique $\mathbb{F}$-optional locally integrable variation process $(V_t^o, t \geq 0)$, called the **$\mathbb{F}$-dual optional projection** *of V such that*

$$\mathbb{E}\left[\int_{[0,\infty)} H_s dV_s\right] = \mathbb{E}\left[\int_{[0,\infty)} H_s dV_s^o\right] \tag{1.10}$$

for any $\mathbb{F}$-optional process H such that $\mathbb{E}\left[\int_{[0,\infty)} |H_s||dV_s|\right] < \infty$.
(b) There exists a unique $\mathbb{F}$-predictable locally integrable variation process $(V_t^p, t \geq 0)$, called the **$\mathbb{F}$-dual predictable projection** *of V such that*

$$\mathbb{E}\left[\int_{[0,\infty)} H_s dV_s\right] = \mathbb{E}\left[\int_{[0,\infty)} H_s dV_s^p\right] \tag{1.11}$$

for any $\mathbb{F}$-predictable process H such that $\mathbb{E}\left[\int_{[0,\infty)} |H_s||dV_s|\right] < \infty$.

Remark 1.35 Note that with our convention that $\Delta V_0 = V_0$, Eq. (1.10) implies that $V_0^o = \mathbb{E}[V_0|\mathscr{F}_0]$. Analogously $V_0^p = \mathbb{E}[V_0|\mathscr{F}_0]$ since $H = 1\!\!1_{[0]} 1\!\!1_F$ for $F \in \mathscr{F}_0$ is an $\mathbb{F}$-predictable process.

Note that predictable processes with finite variation are locally integrable. The terminology *dual predictable projection* refers to the equality

$$\mathbb{E}\left[\int_{[0,\infty)} H_s dV_s^p\right] = \mathbb{E}\left[\int_{[0,\infty)} {}^pH_s dV_s\right]$$

which holds since, by (1.9), $\mathbb{E}\left[\int_{[0,\infty)} H_s dV_s^p\right] = \mathbb{E}\left[\int_{[0,\infty)} {}^pH_s dV_s^p\right]$ and from the definition of dual predictable projection, one has $\mathbb{E}\left[\int_{[0,\infty)} {}^pH_s dV_s^p\right] = \mathbb{E}\left[\int_{[0,\infty)} {}^pH_s dV_s\right]$.

In a general setting, the predictable (resp. optional) projection of an increasing process V is a submartingale whereas the dual predictable (resp. optional) projection is an increasing process.

Proposition 1.36 *Let V be a locally integrable variation process. Then*
(a) ${}^oV - V^o$ and ${}^oV - V^p$ are martingales;
(b) $\Delta V^o = {}^o(\Delta V)$ and $\Delta V^p = {}^p(\Delta V)$.
In particular, if V is càdlàg $\mathbb{F}$-adapted, then V^p is characterized by the fact that $V - V^p$ is an $\mathbb{F}$-martingale.[4]
(c) If V is optional (resp. predictable), $(X \cdot V)^o = {}^oX \cdot V$ (resp. $(X \cdot V)^p = {}^pX \cdot V$).

Proof (a) Firstly, notice that, from the definitions of dual optional and dual predictable projections applied to $H = 1\!\!1_F 1\!\!1_{(s,t]}$ for any $s < t$ and $F \in \mathscr{F}_s$, we get

$$\mathbb{E}[1\!\!1_F(V_t^o - V_s^o)] = \mathbb{E}[1\!\!1_F(V_t - V_s)] = \mathbb{E}[1\!\!1_F(V_t^p - V_s^p)].$$

Secondly, from (1.7) we also get $\mathbb{E}[1\!\!1_F({}^oV_t - {}^oV_s)] = \mathbb{E}[1\!\!1_F(V_t - V_s)]$, which completes the proof.
(b) This is directly implied by definitions of (dual) projections since, for any stopping time T and any set $F \in \mathscr{F}_T$, by taking $H = 1\!\!1_F 1\!\!1_{[\![T]\!]}$, we obtain

$$\mathbb{E}[1\!\!1_F \Delta V_T^o] = \mathbb{E}\left[\int_{[0,\infty)} H_t dV_t^o\right] = \mathbb{E}\left[\int_{[0,\infty)} H_t dV_t\right] = \mathbb{E}[1\!\!1_F \Delta V_T] = \mathbb{E}[1\!\!1_F\, {}^o(\Delta V)_T].$$

The proof for the predictable case is analogous.
(c) For the optional case this is an immediate consequence of (1.8) and (1.10) and for the predictable case of (1.9) and (1.11). □

Remark 1.37 For two $\mathbb{F}$-semimartingales X, Y, the $\mathbb{F}$-predictable covariation process $\langle X, Y\rangle^{\mathbb{F}}$ is (if it exists) the $\mathbb{F}$-dual predictable projection of the covariation process

[4] In other terms, V^p is the compensator of V, with $V_{0-}^p = 0$.

$[X, Y]$. The predictable covariation process exists if the covariation process is locally integrable. The predictable covariation process depends on the filtration.

Example 1.38 Let X and Y be two locally square integrable $\mathbb{F}$-martingales. Then $\langle X, Y\rangle^{\mathbb{F}}$ exists. Indeed, let (τ_n) be a localizing sequence for X and Y. The Kunita–Watanabe inequality (see [189, Corollary after Theorem 25, Chap. II]) implies that $\mathbb{E}\left[\int_{[0,\infty)} 1\!\!1_{[\![0,\tau_n]\!]}|d[X,Y]_s|\right] \leq \sqrt{\mathbb{E}\left[[X]_{\tau_n}\right]}\sqrt{\mathbb{E}\left[[Y]_{\tau_n}\right]}$ for any n. And, since for any square integrable martingale $\mathbb{E}[[X]_{\tau_n}] < \infty$ for each n (see [189, Corollary 3, Chap. II]), we conclude that $[X, Y]$ is a locally integrable variation process.

1.3.3 *Stopping Times and Random Times Decompositions*

In this subsection, we present the decomposition of stopping times into accessible and totally inaccessible parts (see Sect. 1.1.2) and the decomposition of random times into thin and thick parts (defined below, see [9] for a detailed study). To prove these decomposition results one relies on dual predictable and dual optional projections.

Definition 1.39 Let τ be a random time and $A := 1\!\!1_{[\![\tau,\infty[\![}$.
(a) The $\mathbb{F}$-dual optional projection of τ is the $\mathbb{F}$-dual optional projection A^o of A and satisfies, for any non-negative bounded $\mathbb{F}$-optional process Y,

$$\mathbb{E}[Y_\tau 1\!\!1_{\{\tau<\infty\}}] = \mathbb{E}\left[\int_{[0,\infty)} Y_s dA_s^o\right].$$

(b) The $\mathbb{F}$-dual predictable projection of τ is the $\mathbb{F}$-dual predictable projection A^p of A and satisfies, for any non-negative bounded $\mathbb{F}$-predictable process Y,

$$\mathbb{E}[Y_\tau 1\!\!1_{\{\tau<\infty\}}] = \mathbb{E}\left[\int_{[0,\infty)} Y_s dA_s^p\right]. \quad (1.12)$$

If necessary, to avoid possible confusion, we will use the notation $A^{\tau,\mathbb{F},p}$ and $A^{\tau,\mathbb{F},o}$, or $A^{\mathbb{F},p}$ and $A^{\mathbb{F},o}$. If τ is an $\mathbb{F}$-stopping time, then the process $A - A^{\tau,\mathbb{F},p}$ is an $\mathbb{F}$-martingale and $A^{\tau,\mathbb{F},p}$ is the $\mathbb{F}$-compensator of A. If τ is an $\mathbb{F}$-predictable stopping time, then $A^{\tau,\mathbb{F},p} = A$.

The following definition identifies two classes of random times using a criterion based on $\mathbb{F}$-stopping times.

Definition 1.40 A random time τ is called
(a) an $\mathbb{F}$-thick time if $[\![\tau]\!] \cap [\![T]\!]$ is evanescent for any $\mathbb{F}$-stopping time T, i.e., if it avoids all $\mathbb{F}$-stopping times.
(b) an $\mathbb{F}$-thin time if its graph $[\![\tau]\!]$ is contained in a thin set, i.e., if there exists a sequence of $\mathbb{F}$-stopping times $(T_n)_{n=1}^\infty$ with disjoint graphs such that $[\![\tau]\!] \subset \bigcup_n [\![T_n]\!]$. We say that such a sequence $(T_n)_n$ exhausts the thin time τ or that $(T_n)_n$ is an exhausting sequence of the thin time τ.

Concerning the above definition, let us remark that an exhausting sequence $(T_n)_n$ of a thin time is not unique.

The straightforward observation, that the sets of thick and thin times have trivial intersection, is stated in the following lemma. The same property holds for accessible and totally inaccessible stopping times.

Lemma 1.41 *(a) A stopping time T belongs to the class of accessible and totally inaccessible stopping times if and only if $T = \infty$.*
(b) A random time τ belongs to the class of thick times and to the class of thin times if and only if $\tau = \infty$.

In the following theorem, we present the main object of interest of this subsection, namely, two decompositions of stopping times and random times respectively.

Theorem 1.42 *(a) Any $\mathbb{F}$-stopping time T has a decomposition (T_1, T_2) such that T_1 is an $\mathbb{F}$-accessible stopping time, T_2 is an $\mathbb{F}$-totally inaccessible stopping time and*

$$T_1 \wedge T_2 = T \quad \textit{and} \quad T_1 \vee T_2 = \infty.$$

This decomposition is unique on the set $\{T < \infty\}$.
(b) Any random time τ has a decomposition (τ_1, τ_2) such that τ_1 is a thick time, τ_2 is a thin time, and

$$\tau = \tau_1 \wedge \tau_2 \quad \textit{and} \quad \tau_1 \vee \tau_2 = \infty.$$

This decomposition is unique on the set $\{\tau < \infty\}$.

Proof We prove (b) the proof of (a) is analogous, using A^p instead of A^o.
(b) Take τ_1 and τ_2 of the following form $\tau_1 = \tau_{\{\Delta A^o_\tau = 0\}}$ and $\tau_2 = \tau_{\{\Delta A^o_\tau > 0\}}$, where the definition of the restriction of a random time is given in (1.1). Properties of dual optional projection ensure that τ_1 and τ_2 satisfy the required conditions. More precisely, τ_1 is a thick time as, for any $\mathbb{F}$-stopping time T,

$$\begin{aligned}\mathbb{P}(\tau_1 = T < \infty) &= \mathbb{E}\left[\mathbb{1}_{\{\tau = T\}\cap\{\Delta A^o_\tau = 0\}}\mathbb{1}_{(T<\infty)}\right] \\ &= \mathbb{E}\left[\int_{[0,\infty)} \mathbb{1}_{\{u=T\}\cap\{\Delta A^o_u = 0\}} dA^o_u\right] = 0\end{aligned}$$

and the time τ_2 is a thin time as

$$[\![\tau_2]\!] = [\![\tau]\!] \cap \{\Delta A^o > 0\} = [\![\tau]\!] \cap \bigcup_n [\![T_n]\!] \subset \bigcup_n [\![T_n]\!],$$

where the sequence $(T_n)_n$ exhausts the jumps of the càdlàg increasing process A^o, i.e., $\{\Delta A^o > 0\} = \bigcup_n [\![T_n]\!]$. □

The next result gives some equivalent characterisations of different classes of stopping times and random times.

Theorem 1.43 *(a) An $\mathbb{F}$-stopping time is an $\mathbb{F}$-totally inaccessible stopping time if and only if its $\mathbb{F}$-dual predictable projection is a continuous process.*

An $\mathbb{F}$-stopping time is an $\mathbb{F}$-accessible stopping time if and only if its $\mathbb{F}$-dual predictable projection is a pure jump process.

(b) A random time is an $\mathbb{F}$-thick time if and only if its $\mathbb{F}$-dual optional projection is a continuous process.

A random time is an $\mathbb{F}$-thin time if and only if its $\mathbb{F}$-dual optional projection is a pure jump process.

Proof As for the previous result, the proofs of assertions (a) and (b) are analogous, thus we provide only one.

(b) Let τ be a random time and T be an $\mathbb{F}$-stopping time. Since A^o is an increasing process and $\mathbb{E}[\Delta A^o_T \mathbb{1}_{\{T<\infty\}}] = \mathbb{P}(\tau = T < \infty)$, we deduce that

$$\mathbb{P}(\tau = T < \infty) = 0 \quad \text{if and only if} \quad \Delta A^o_T \mathbb{1}_{\{T<\infty\}} = 0 \quad \mathbb{P}\text{-a.s.}$$

Since $\{\Delta A^o > 0\}$ is an optional set, the optional section theorem (Theorem 1.7 (a)) implies that $\{\Delta A^o > 0\}$ is exhausted by disjoint graphs of $\mathbb{F}$-stopping times. Thus, we conclude that τ is a thick time if and only if A^o is continuous.

For $(T_n)_n$ a sequence of $\mathbb{F}$-stopping times with disjoint graphs, we have

$$\sum_n \mathbb{P}(\tau = T_n < \infty) = \sum_n \mathbb{E}[\Delta A^o_{T_n} \mathbb{1}_{\{T_n<\infty\}}].$$

Since $\mathbb{E}[A^o_\infty] = \mathbb{P}(\tau < \infty)$, by definition of the dual optional projection and using the fact that A^o is an increasing process, we conclude that the sequence $(T_n)_n$ satisfies the condition $\sum_n \mathbb{P}(\tau = T_n < \infty) = \mathbb{P}(\tau < \infty)$ if and only if it satisfies the condition $\mathbb{E}[A^o_\infty] = \sum_n \mathbb{E}\left[\Delta A^o_{T_n} \mathbb{1}_{\{T_n<\infty\}}\right]$. In other words, τ is a thin time if and only if A^o is a pure jump process. □

1.3.4 Azéma's Supermartingales

Definition 1.44 Let τ be a random time and $\mathbb{F}$ be a filtration satisfying the usual conditions.

(a) The **default indicator process** is defined by $A := \mathbb{1}_{[\![\tau,\infty[\![}$ and its natural filtration by $\mathbb{A} := (\mathscr{A}_t, t \geq 0)$.

(b) The **Azéma supermartingales** Z and $\widetilde{Z}$ are defined by $Z := {}^o(1 - A)$ and $\widetilde{Z} := {}^o(1 - A_-)$, i.e., $Z_t = \mathbb{P}(\tau > t|\mathscr{F}_t)$ and $\widetilde{Z}_t = \mathbb{P}(\tau \geq t|\mathscr{F}_t)$.

The supermartingale Z is a càdlàg process while the supermartingale $\widetilde{Z}$ admits right- and left-limits only. Note that $\widetilde{Z}_+ = Z_+ = Z$ and $\widetilde{Z}_- = Z_-$.

Lemma 1.45 *The processes Z and $\widetilde{Z}$ are supermartingales of class (D).*

Proof Being bounded, Z is of class (D). The supermartingale property follows from the fact that, for $s < t$, by the tower property of conditional expectations

$$\mathbb{E}[Z_t|\mathscr{F}_s] = \mathbb{E}[\mathbb{1}_{\{\tau>t\}}|\mathscr{F}_s] \le \mathbb{E}[\mathbb{1}_{\{\tau>s\}}|\mathscr{F}_s] = Z_s .$$

□

Proposition 1.46 *We study the processes introduced in Definition 1.44.*
(a) The Doob–Meyer decomposition of the supermartingale Z is

$$Z_t = \mathbb{E}\left[A^p_\infty + \mathbb{1}_{\{\tau=\infty\}}|\mathscr{F}_t\right] - A^p_t = n_t - A^p_t \tag{1.13}$$

where $n_t := \mathbb{E}\left[A^p_\infty + \mathbb{1}_{\{\tau=\infty\}}|\mathscr{F}_t\right]$ is a BMO-martingale and $n_0 = 1$.
(b) The supermartingale Z has the following optional decomposition

$$Z_t = \mathbb{E}\left[A^o_\infty + \mathbb{1}_{\{\tau=\infty\}}|\mathscr{F}_t\right] - A^o_t = m_t - A^o_t \tag{1.14}$$

where $m_t := \mathbb{E}\left[A^o_\infty + \mathbb{1}_{\{\tau=\infty\}}|\mathscr{F}_t\right]$ is a BMO-martingale and $m_0 = 1$.
(c) The supermartingale $\widetilde{Z}$ has a decomposition

$$\widetilde{Z} = m - A^o_- . \tag{1.15}$$

(d) The following relations hold : $\widetilde{Z} = Z_- + \Delta m$ and $\widetilde{Z} = Z + \Delta A^o$.

Proof (a) Let t be fixed and let $F \in \mathscr{F}_t$. Then, the càg process $Y = \mathbb{1}_F \mathbb{1}_{(t,\infty)}$ is $\mathbb{F}$-predictable. From (1.12),

$$\mathbb{E}\left[Y_\tau \mathbb{1}_{\{\tau<\infty\}}\right] = \mathbb{E}\left[\mathbb{1}_F \mathbb{1}_{\{t<\tau<\infty\}}\right] = \mathbb{E}\left[\mathbb{1}_F(A^p_\infty - A^p_t)\right].$$

It follows that $n_t = \mathbb{E}[A^p_\infty + \mathbb{1}_{\{\tau=\infty\}}|\mathscr{F}_t] = Z_t + A^p_t$, and the process n is a non-negative martingale. It remains to show that n is a BMO-martingale. Let T be an $\mathbb{F}$-stopping time. Then, since A^p is increasing,

$$|n_\infty - n_{T-}| \le |n_\infty - A^p_T| + |n_{T-} - A^p_T| = \mathbb{P}(\tau = \infty|\mathscr{F}_\infty) + A^p_\infty - A^p_T + |n_{T-} - A^p_T|$$

so by taking conditional expectations and using the definition of dual predictable projection, we obtain

$$\mathbb{E}\left[|n_\infty - n_{T-}||\mathscr{F}_T\right] \le \mathbb{E}\left[\mathbb{1}_{\{T<\tau\le\infty\}}|\mathscr{F}_T\right] + \mathbb{E}\left[|n_{T-} - A^p_T||\mathscr{F}_T\right]. \tag{1.16}$$

Moreover, for any $\mathbb{F}$-predictable stopping time S, we have on $\{S < \infty\}$

$$n_{S-} - A^p_S = \mathbb{E}\left[n_\infty - A^p_S|\mathscr{F}_{S-}\right] = \mathbb{E}\left[\mathbb{1}_{\{S<\tau\le\infty\}}|\mathscr{F}_{S-}\right] \in [0, 1] \tag{1.17}$$

which, by the predictable section theorem and using the fact that $|n_- - A^p|$ is a predictable process, implies that $|n_- - A^p| \le 1$ a.s. Combining (1.16) and (1.17)

we obtain that $\|n\|_{BMO_1} \leq 2$. Using the fact that $Z_0 = 1 - \mathbb{P}(\tau = 0|\mathcal{F}_0) = 1 - A_0^p$, it follows that $n_0 := \mathbb{E}\left[A_\infty^p + 1\!\!1_{\{\tau=\infty\}}|\mathcal{F}_0\right] = 1$. The proof of (b) is analogous.
(c) For an $\mathbb{F}$-stopping time T, applying the definition of dual optional projection to the optional process $Y = 1\!\!1_F 1\!\!1_{[\![T,\infty[\![}$ for $F \in \mathcal{F}_T$, leads to $\widetilde{Z}_T = m_T - A_{T-}^o$ and we conclude by the optional section theorem. The proof of (d) is a straightforward application of the previous assertions. □

In particular Proposition 1.46 (c) implies that Z_- is the $\mathbb{F}$-predictable projection of $1 - A_- = 1\!\!1_{[\![0,\tau]\!]}$. Indeed, using (1.15), we see that

$$^p(1 - A_-) = {}^p(\widetilde{Z}) = {}^pm - A_-^o = m_- - A_-^o = Z_- \tag{1.18}$$

since the predictable projection of a u.i. martingale is its left-limit (Example 1.33).

Definition 1.47 We introduce two frequently used conditions:
Condition (**A**): the random time τ **a**voids $\mathbb{F}$-stopping times, i.e., for any $\mathbb{F}$-stopping time ϑ, $\mathbb{P}(\tau = \vartheta < \infty) = 0$, or equivalently τ is a thick time.
Condition (**C**): all $\mathbb{F}$-martingales are **c**ontinuous.

Lemma 1.48 *Let τ be a random time.*
(a) Under condition (**A**), *$A^p = A^o$ and these processes are continuous.*
(b) Under conditions (**C**) *and* (**A**), *Z is continuous.*
(c) Under condition (**C**), *$A^p = A^o$.*

Proof (a) Assume that (**A**) holds. If ϑ is a jump time of A^p, it is an $\mathbb{F}$-predictable stopping time and $\mathbb{E}\left[A_\vartheta^p - A_{\vartheta-}^p\right] = \mathbb{E}[1\!\!1_{\{\tau=\vartheta\}}] = 0$; the continuity of A^p follows.
(b) follows from (a) and the condition (**C**).
(c) Under (**C**), the predictable and optional σ-fields are equal, hence A^o is predictable. □

Note that if one of the conditions (**C**) or (**A**) is satisfied, then $Z = \widetilde{Z}$.

Proposition 1.49 *Let m and n be the $\mathbb{F}$-BMO martingales defined in Proposition 1.46 and Y be an $H^1(\mathbb{F})$-martingale. Then*
(a) the processes $[Y, m]$ and $[Y, n]$ have an integrable variation;
(b) $\mathbb{E}[Y_\tau] = \mathbb{E}[[Y, m]_\infty] = \mathbb{E}[[Y, n]_\infty] + \mathbb{E}[(\Delta Y \cdot A^o)_\infty]$.

Proof (a) Let Y be an $\mathbb{F}$-martingale of the class H^1. Then, since m is an $BMO(\mathbb{F})$-martingale as proven in Proposition 1.46, Fefferman's inequality (1.5) implies that $[Y, m]$ has integrable variation. Same arguments apply to $[Y, n]$.
(b) By the integration by parts formula given in Proposition 1.16, the optional decomposition of Z given in (1.14) and $m_0 = 1$, we have

$$Y_\infty Z_\infty = \int_0^\infty Z_{s-} dY_s + \int_0^\infty Y_{s-} dm_s - \int_{[0,\infty)} Y_s dA_s^o + [m, Y]_\infty.$$

Assume now that Y is bounded. Then, by taking expectation of both sides, and using the definition of the dual optional projection, we get

$$\mathbb{E}\left[Y_\infty \mathbb{1}_{\{\tau=\infty\}}\right] = \mathbb{E}\left[[m, Y]_\infty\right] - \mathbb{E}\left[Y_\tau \mathbb{1}_{\{\tau<\infty\}}\right]$$

which shows the assertion for bounded martingales. To extend this result to any $Y \in H^1$ we use the duality (H^1, BMO), the fact that $Y \to \mathbb{E}[Y_\tau]$ is a linear form and that m is a BMO martingale.

To show the second equality we note that, by (1.13) and (1.14), $m = n + A^o - A^p$. Then, by Proposition 1.16 (b) and (c) and Remark 1.35, we obtain that

$$\mathbb{E}\left[[Y, m]_\infty\right] = \mathbb{E}\left[[Y, n]_\infty\right] + \mathbb{E}\left[(\Delta Y \cdot A^o)_\infty\right] - \mathbb{E}\left[(\Delta A^p \cdot Y)_\infty\right].$$

Let T be a predictable stopping time, then, by the properties of dual predictable projection, we have

$$\Delta A^p_T \mathbb{1}_{\{T<\infty\}} = \mathbb{E}\left[\Delta A^p_T \mathbb{1}_{\{T<\infty\}} | \mathscr{F}_{T-}\right] = \mathbb{P}(\tau = T < \infty | \mathscr{F}_{T-}) \in [0, 1].$$

Since A^p is a predictable process, by the predictable section theorem (Theorem 1.7 (b)), we conclude that $\Delta A^p \in [0, 1]$. Thus $\Delta A^p \cdot Y$ is an $\mathbb{F}$-martingale and $\mathbb{E}[\Delta A^p \cdot Y] = 0$. □

Remark 1.50 Let Y be a u.i. $\mathbb{F}$-martingale. Then $\mathbb{E}[Y_\tau] = \mathbb{E}[Y_\infty m_\infty]$. Indeed,

$$\mathbb{E}[Y_\tau] = \mathbb{E}\left[\int_{[0,\infty)} Y_s dA^o_s\right] + \mathbb{E}[Y_\infty Z_\infty] = \mathbb{E}[Y_\infty A^o_\infty] + \mathbb{E}[Y_\infty Z_\infty] = \mathbb{E}[Y_\infty m_\infty]$$

where the second equality comes from the integration by parts formula (1.2).

Lemma 1.51 *Let $R := \inf\{t : Z_t = 0\}$. Then, $R := \inf\{t : Z_{t-} = 0\} = \inf\{t : \widetilde{Z}_t = 0\}$ and $\tau \leq R$. The $\mathbb{F}$-stopping time R is the smallest $\mathbb{F}$-stopping time greater than τ.*

Proof First we show that $\{Z_t = 0\} = \{\forall \varepsilon > 0, \widetilde{Z}_{t+\varepsilon} = 0\}$. Inclusion "$\supset$" comes from $\widetilde{Z} \geq Z$ and right-continuity of Z. To prove the inclusion "$\subset$", we note that

$$\begin{aligned} 0 = \mathbb{E}\left[\mathbb{1}_{\{Z_t=0\}} Z_{t+\varepsilon}\right] &= \mathbb{E}\left[\mathbb{1}_{\{Z_t=0\}} \mathbb{1}_{\{\tau>t+\varepsilon\}}\right] \\ &\geq \mathbb{E}\left[\mathbb{1}_{\{Z_t=0\}} \mathbb{1}_{\{\tau=t+2\varepsilon\}}\right] = \mathbb{E}\left[\mathbb{1}_{\{Z_t=0\}} \Delta A^o_{t+2\varepsilon}\right]. \end{aligned}$$

Then the result comes from the relation $\widetilde{Z} = Z + \Delta A^o$. The stopping time R is the smallest stopping time greater than τ as for any other stopping time T such that $T \geq \tau$, we have $Z_T = 0$. □

1.3.5 Properties of Random Sets

In this subsection, we present some technical (and useful) results on random sets.

A random set $C \subset \Omega \times \mathbb{R}^+$ is a **closed** set if for each ω, the ω-section set

$$C(\omega) := \{t : (\omega, t) \in C\}$$

is closed in $\mathbb{R}^+$. A random set C is a left-closed set if for each ω, the ω-section set $C(\omega)$ is closed in $\mathbb{R}^+$ for upper limit topology, i.e., the limit of any *increasing* converging sequence of elements from C belongs to C (in particular, the set $(a, b]$ for $a < b$ is left-closed).

Let V be a non-negative increasing process. Recall that, by Definition 1.3, V is in particular càdlàg. Then, we note by $S(V)$ and $S^\ell(V)$ respectively the **support** and **left-support** of V, defined as

$$S(V) = \{t : \forall \varepsilon > 0, \ \ V_t - V_{t-\varepsilon} > 0 \ \text{ or } \ V_{t+\varepsilon} - V_t > 0\},$$
$$S^\ell(V) = \{t : \forall \varepsilon > 0, \ \ V_t - V_{t-\varepsilon} > 0\},$$

where $V_t = 0$ for $t < 0$. Note that $\mathbb{P}$-a.s., since V is càdlàg, $S(V)(\omega)\backslash S^\ell(V)(\omega)$ is at most a countable set and for each $t \in S(V)(\omega)\backslash S^\ell(V)(\omega)$ one has $\Delta V_t(\omega) = 0$. Therefore, $dV(\omega) = \mathbb{1}_{S^\ell(V)} dV(\omega) = \mathbb{1}_{S(V)} dV(\omega)$ $\mathbb{P}$-a.s.

Lemma 1.52 *Let V be an increasing integrable process, V^o (resp. V^p) its dual optional (resp. predictable) projection. Then, $S(V) \subset S(V^o)$, $S^\ell(V) \subset S^\ell(V^o)$ and $S^\ell(V) \subset S^\ell(V^p)$. Furthermore $S^\ell(V^o)$ (resp. $S^\ell(V^p)$) is the smallest optional (resp. predictable) set containing $S^\ell(V)$.*

Proof Consider the optional process $H = \mathbb{1}_{\{S^\ell(V^o)\}^c}$. By the definition of the dual optional projection

$$\mathbb{E}\left[\int_{[0,\infty)} H_u dV_u\right] = \mathbb{E}\left[\int_{[0,\infty)} H_u dV_u^o\right] = 0$$

which shows that $S^\ell(V) \subset S^\ell(V^o)$ and $S(V) = \overline{S^\ell(V)} \subset \overline{S^\ell(V^o)} = S(V^o)$. For the remaining inclusion, we use a similar argument by noting that the process $\mathbb{1}_{\{S^\ell(V^p)\}^c}$ is predictable. It remains to prove that $S^\ell(V^o)$ is the smallest optional set containing $S^\ell(V)$. Let Γ be an optional set containing $S^\ell(V)$. Then

$$\mathbb{E}\left[\int_{[0,\infty)} \mathbb{1}_\Gamma(s) dV_s^o\right] = \mathbb{E}\left[\int_{[0,\infty)} \mathbb{1}_\Gamma(s) dV_s\right] = \mathbb{E}[V_\infty] = \mathbb{E}[V_\infty^o]$$

which shows that $S^\ell(V^o) \subset \Gamma$. The predictable case is solved in a similar way. □

Lemma 1.53 *Let C be a left-closed (resp. closed) measurable set. Then*
(a) The left-closed (resp. closed) set $\{{}^o(\mathbb{1}_C) = 1\}$ is the biggest optional set included in C.
(b) The left-closed set $\{{}^p(\mathbb{1}_C) = 1\}$ is the biggest predictable set included in C.

Proof Let C is a left-closed set and N be an optional set such that $N \subset C$. Then $\mathbb{1}_N \leq \mathbb{1}_C$ and, by taking optional projection of the both sides, $\mathbb{1}_N \leq {}^o(\mathbb{1}_C)$. Therefore, we conclude that $N \subset \{{}^o(\mathbb{1}_C) = 1\}$. It remains to show that $\{{}^o(\mathbb{1}_C) = 1\} \subset C$. Assume that $\{{}^o(\mathbb{1}_C) = 1\} \cap C^c$ is not evanescent, i.e., there exists a random time τ such that $[\![\tau]\!] \subset \{{}^o(\mathbb{1}_C) = 1\} \cap C^c$ and $\mathbb{P}(\tau < \infty) > 0$. Define an increasing process $A = \mathbb{1}_{[\![\tau,\infty[\![}$. Since $S^\ell(A) = [\![\tau]\!] \subset \{{}^o(\mathbb{1}_C) = 1\}$ we have that $A = {}^o(\mathbb{1}_C) \cdot A$ and

$$(\mathbb{1}_{C^c} \cdot A^o)^o = {}^o(\mathbb{1}_{C^c}) \cdot A^o = ({}^o(\mathbb{1}_{C^c}) \cdot A)^o = 0.$$

Thus, $S^\ell(A) \subset S^\ell(A^o) \subset C$, which is a contradiction. □

Lemmas 1.52 and 1.53 can be applied to the default indicator process of a random time resulting in the following proposition.

Proposition 1.54 *Let τ be a random time and processes A, Z, $\widetilde{Z}$, A^o and A^p given in Definitions 1.44 and 1.39. Then the following assertions hold.*

(a) The set $S^\ell(A^o)$ is the smallest optional set containing $[\![\tau]\!]$.
(b) The set $S^\ell(A^p)$ is the smallest predictable set containing $[\![\tau]\!]$.
(c) The set $\{\widetilde{Z} = 1\}$ is the biggest optional set contained in $[\![0, \tau]\!]$.
(d) The set $\{Z_- = 1\}$ is the biggest predictable set contained in $[\![0, \tau]\!]$.

In particular, if τ coincides with the end of an optional (resp. predictable) set then $\widetilde{Z}_\tau = 1$ (resp. $Z_{\tau-} = 1$).

1.4 Essentials of Mathematical Finance

We assume that d risky assets are traded in a financial market with price S which is a d-dimensional $\mathbb{F}$-semimartingale. For simplicity we assume that the savings account is an asset with constant price. A portfolio (or a strategy) is a d-dimensional $\mathbb{F}$-predictable process π, where π is the number of shares of the asset S that the agent holds. The *self-financing condition* means that the process X defined by $X := x + \pi \cdot S$ is the wealth associated to the portfolio π with initial wealth $X_0 = x$. The portfolio is assumed to satisfy $\pi \in L(S, \mathbb{F})$, where $L(S, \mathbb{F})$ is the set of $\mathbb{F}$-predictable processes, stochastically integrable w.r.t. S. The obvious equality $X = (X - \pi S) + \pi S$ means that the agent has invested the amount $(X - \pi S)$ in the savings account and πS in the risky assets.

The time horizon $[0, T\rangle$ is defined as the interval $[0, T]$ if $T \in (0, \infty)$ and as the interval $[0, \infty)$ if $T = \infty$. For $a \in \mathbb{R}^+$, an element $\pi \in L(S, \mathbb{F})$ is said to be an a-admissible $\mathbb{F}$-strategy on $[0, T\rangle$ if $(\pi \cdot S)_t \geq -a$ $\mathbb{P}$-a.s. for all $t \in [0, T\rangle$. We denote by $\mathscr{S}_a(\mathbb{F})$ the set of all a-admissible $\mathbb{F}$-strategies (the choice of the time horizon being clear). A process $\pi \in L(S, \mathbb{F})$ is called an *admissible* $\mathbb{F}$*-strategy* if $\pi \in \mathscr{S}(\mathbb{F}) := \bigcup_{a \in \mathbb{R}^+} \mathscr{S}_a(\mathbb{F})$.

We recall some standard arbitrage conditions. As before, we assume zero interest rate and that the market $(S, \mathbb{F})$ is a market where S and the savings account are traded using $\mathbb{F}$-adapted strategies.

An admissible strategy π yields a classical arbitrage opportunity on the time horizon $[0, T\rangle$ if $(\pi \cdot S)_T \geq 0$ $\mathbb{P}$-a.s. and $\mathbb{P}((\pi \cdot S)_T > 0) > 0$. If there exists no such $\pi \in \mathscr{S}(\mathbb{F})$ we say that the financial market $(S, \mathbb{F})$ satisfies **no arbitrage (NA)** on the time horizon $[0, T\rangle$.

The financial market $(S, \mathbb{F})$ satisfies **no free lunch with vanishing risk** on the time horizon $[0, T\rangle$ (NFLVR($[0, T\rangle$)) if and only if there exists a **equivalent martingale measure** on the time horizon $[0, T\rangle$ (EMM($[0, T\rangle$)), i.e., a probability measure $\mathbb{Q}$, such that $\mathbb{Q}$ is equivalent to $\mathbb{P}$ on $\mathscr{F}_T$ and the process $(S_t)_{t\in[0,T\rangle}$ is a $(\mathbb{Q}, \mathbb{F})$-sigma martingale.[5]

The market $(S, \mathbb{F})$ is **complete** on the time horizon $[0, T\rangle$ if, for any bounded $\mathscr{F}_T$-measurable r.v. H, there exists a hedging strategy, i.e., a pair (x, π) such that $\pi \in \mathscr{S}(\mathbb{F})$ and $x + \pi \cdot S_T = H$. If there exists an EMM in a complete market then this EMM is unique.

The financial market $(S, \mathbb{F})$ satisfies **no unbounded profit with bounded risk** on the time horizon $[0, T\rangle$ (NUPBR($[0, T\rangle$)) if the set

$$\{(\pi \cdot S)_T : \pi \in L(S, \mathbb{F}) \text{ and } (\pi \cdot S)_t \geq -1 \text{ for } t \in [0, T\rangle\}$$

is bounded in probability.

We recall that the financial market $(S, \mathbb{F})$ satisfies NFLVR($[0, T\rangle$) if and only if $(S, \mathbb{F})$ satisfies both NA($[0, T\rangle$) and NUPBR($[0, T\rangle$).

In this book we are particularly interested in the following condition. The financial market $(S, \mathbb{F})$ satisfies **no unbounded profit with bounded risk (NUPBR)** if $(S, \mathbb{F})$ satisfies NUPBR($[0, T]$) for any $T \in (0, \infty)$.

The process S is said to admit a **$\mathbb{F}$-local martingale deflator** if there exists a positive $\mathbb{F}$-local martingale L with $L_0 = 1$ such that the process LS is an $\mathbb{F}$-local martingale. If there exists a local martingale deflator, then NUPBR holds.

The process S is said to admit a **supermartingale deflator** if there exists a positive $\mathbb{F}$-supermartingale Y such that $Y(1+\pi \cdot S)$ is a supermartingale for any $\pi \in L(S, \mathbb{F})$ such that $(\pi \cdot S)_t \geq -1$ for all $t < \infty$. The existence of a supermartingale deflator is equivalent to NUPBR.

1.5 Bibliographic Notes

All the results on stochastic processes which we have recalled without proofs can be found in Cohen and Elliott [63], Dellacherie [72], Dellacherie and Meyer [77, 79],

[5] A process X is a sigma-martingale if there exists a martingale Y and an Y-integrable predictable non-negative process φ such that $X = \varphi \cdot Y$. A local martingale is a sigma-martingale.

He et al. [112], Jacod and Shiryaev [125], Protter [189], Revuz and Yor [190] and Rogers and Williams [192].

Right-continuity of filtration is an important property and is a key tool to prove, e.g., existence of càdlàg modification of a martingale, Doob–Meyer decomposition of a supermartingale, see He et al. [112, Chap. 2, Sect. 5] for more information. The lack of this property for the case of a filtration generated by two right-continuous filtrations is presented in Williams [209, p. 48] and Weizsacker [208].

The study of BMO spaces can be found in Dellacherie and Meyer [79, Chap. VII]. The Bichteler–Dellacherie–Mokobodzki theorem was proven independently by Bichteler [33, 34] and Dellacherie [73]. The Doob–Meyer decomposition of a supermartingale which is not càdlàg is studied in Mertens [176]. More information on the Azéma supermartingales is presented in Azéma [24] and Jeulin and Yor [141]. See also Chaps. 2 and 5 in this book. The result from Proposition 1.24 (a) does not extend to local martingales, see the related studies Stricker [203], Föllmer and Protter [96] and Larsson [165]. Thin times and random times decomposition are introduced and studied in Aksamit et al. [9].

No Free Lunch with Vanishing Risk (NFLVR) is presented in detail in Delbaen and Schachermayer [70] and in Kabanov et al. [148]. Fontana summarizes in [98] various arbitrage conditions, Kardaras [151], Karatzas and Kardaras [149] and Takaoka and Schweizer [206] give results on characterisation of NUPBR via deflators. In particular NUPBR is equivalent to no arbitrage of the first kind (NA1).

1.6 Exercises

Exercise 1.1 Show that for an $\mathbb{F}$-stopping time τ, one has $\tau \in \mathscr{F}_{\tau-}$ and $\mathscr{F}_{\tau-} \subset \mathscr{F}_{\tau}$. Find an example where $\mathscr{F}_{\tau-} \neq \mathscr{F}_{\tau}$.

Exercise 1.2 Give an example where τ is a $\mathbb{G}$-stopping time but not an $\mathbb{F}$-stopping time, where $\mathbb{F} \subset \mathbb{G}$. Give an example where τ is a $\mathbb{G}$-predictable stopping time and an $\mathbb{F}$-stopping time, but not a predictable $\mathbb{F}$-stopping time.

Exercise 1.3 Let B be a Brownian motion. Prove that $W := \int_0^\cdot \mathrm{sgn}(B_s)dB_s$ defines an $\mathbb{F}^B$ and an $\mathbb{F}^W$-Brownian motion. Prove that $\beta := B - \int_0^\cdot \frac{B_s}{s} ds$ is a Brownian motion in its natural filtration which is not an $\mathbb{F}^B$-Brownian motion.

Exercise 1.4 Let X be a measurable process such that $\mathbb{E}[\int_0^t |X_s|ds] < \infty$ for each $t \geq 0$. Define $Y = \int_0^\cdot X_s ds$. Prove that $M := {}^oY - \int_0^\cdot {}^oX_s ds$ is an $\mathbb{F}$-martingale. In particular, for any (bounded) not necessarily $\mathbb{F}$-adapted process a, the process $\mathbb{E}\left[\int_0^t a_u du | \mathscr{F}_t\right] - \int_0^t \mathbb{E}[a_u|\mathscr{F}_u]du$ is an $\mathbb{F}$-martingale.

Exercise 1.5 Give an example of a random time τ where A^p and A^o are different.

Exercise 1.6 (*Doob's Maximal Identity*) Let X be a non-negative continuous local martingale such that $X_0 = x > 0$ and $\lim_{t\to\infty} X_t = 0$. Prove that $\mathbb{P}\left(\sup_{t\geq 0} X_t > a\right) = \left(\frac{x}{a}\right) \wedge 1$ and $\sup_{t\geq 0} X_t \overset{\text{law}}{=} \frac{x}{U}$ where U is a random variable with a uniform law on $[0, 1]$.

Chapter 2
Compensators of Random Times

Given a random time τ, we study the compensator of the default (indicator) process $A := \mathbb{1}_{[\![\tau,\infty[\![}$ and the associated compensated martingale M. As a financial application, we establish properties of the intensity rate for a single default, and give pricing formulae for defaultable derivatives.

2.1 Compensator of a Default Process in Its Natural Filtration

Let τ be a random time on a probability space $(\Omega, \mathscr{F}, \mathbb{P})$. We work with the default indicator process $A := \mathbb{1}_{[\![\tau,\infty[\![}$ and its natural filtration $\mathbb{A} := (\mathscr{A}_t, t \geq 0)$, both introduced in Definition 1.44. Equivalently, the filtration $\mathbb{A}$ is the smallest filtration satisfying the usual conditions, which renders τ a stopping time. It is important (and obvious) to note that $\int_0^t g(u)dA_u := \int_{(0,t]} g(u)dA_u = (A_t - A_0)g(\tau)$ where g is a Borel function. We denote by F the right-continuous cumulative distribution function of τ, defined as $F(t) = \mathbb{P}(\tau \leq t)$ for any $t \in \mathbb{R}$ and $F(\infty) = \lim_{t\to\infty} F(t) = 1 - \mathbb{P}(\tau = \infty)$. Note that, unless τ is $\mathbb{P}$-a.s. finite, $F(\infty)$ is not equal to one. As $\Delta F(t) = F(t) - F(t-) = \mathbb{P}(\tau = t)$, τ is an $\mathbb{A}$-totally inaccessible stopping time if and only if F is continuous (see [77, Chap. IV]). Remark as well that the càg process A_- is $\mathbb{A}$-predictable.

Lemma 2.1 *The filtration $\mathbb{A}$ satisfies $\mathscr{A}_t = \sigma(\{\tau \leq s\} : s \leq t) \vee \mathscr{N}^{\mathbb{P}}$, where $\mathscr{N}^{\mathbb{P}}$ is the set of $\mathbb{P}$-null sets.*

Proof Let us introduce an auxiliary filtration $\mathbb{A}^0 := (\mathscr{A}_t^0 : t \geq 0)$ defined by

$$\mathscr{A}_t^0 := \sigma(\{\tau \leq s\} : s \leq t) \vee \mathscr{N}^{\mathbb{P}}.$$

It is then enough to show that $\mathbb{A}^0$ is right-continuous. Since $\mathscr{A}_\infty^0 = \sigma(\tau)$ any element of $\mathscr{A}_t^0$ can be written as $\{\tau \in B\}$ for $B \in \mathscr{B}(\mathbb{R}^+)$ such that $B = B \cap [0, t]$. Since for

A. Aksamit and M. Jeanblanc, *Enlargement of Filtration with Finance in View*, SpringerBriefs in Quantitative Finance,
https://doi.org/10.1007/978-3-319-41255-9_2

any $\{\tau \in B\} \in \mathscr{A}^0_{t+}$, the Borel set B satisfies

$$B = \bigcap_{r>t} B \cap [0, r] = B \cap [0, t]$$

we deduce that $\mathscr{A}^0_{t+} = \mathscr{A}^0_t$. □

We present an important lemma, which explains the structure of $\mathscr{A}_t$-measurable random variables.

Lemma 2.2 *A random variable Y is $\mathscr{A}_t$-measurable if and only if it is of the form $Y = h\mathbb{1}_{\{t<\tau\}} + g(\tau)\mathbb{1}_{\{\tau\leq t\}}$ a.s. where h is a constant and g is a Borel function.*

In particular, if $A \in \mathscr{A}_t$, then $A \cap \{t < \tau\} = \widehat{A} \cap \{t < \tau\}$ for some $\widehat{A} \in \mathscr{A}_0$ (note that $\mathbb{P}(\widehat{A}) = 0$ or $\mathbb{P}(\widehat{A}) = 1$).

Proof Let $\mathscr{V}$ be the vector space of random variables defined as follows

$$\mathscr{V} := \{Y = h\mathbb{1}_{\{t<\tau\}} + g(\tau)\mathbb{1}_{\{\tau\leq t\}} : h \in \mathbb{R},\ \ g : \overline{\mathbb{R}}^+ \to \mathbb{R} \text{ is a Borel function}\}$$

Any random variable $Y \in \mathscr{V}$ is $\mathscr{A}_t$-measurable since, for any $B \in \mathscr{B}(\mathbb{R})$, we have

$$Y^{-1}(B) = \{\omega : h \in B,\ t < \tau(\omega)\} \cup \{\omega : g(\tau(\omega) \wedge t) \in B, \tau(\omega) \leq t\} \in \mathscr{A}_t$$

as $\tau \wedge t$ is an $\mathscr{A}_t$-measurable random variable. On the other hand, since $\mathscr{V}$ satisfies

1. $1 \in \mathscr{V}$,
2. if $Y^n := h^n\mathbb{1}_{\{t<\tau\}} + g^n(\tau)\mathbb{1}_{\{\tau\leq t\}}$ where, for each n, h^n is a constant and g^n is a Borel function, $Y^n \nearrow Y$ and Y is finite, then $Y = h\mathbb{1}_{\{t<\tau\}} + g(\tau)\mathbb{1}_{\{\tau\leq t\}}$, where $h := \limsup h^n$ and $g := \limsup g^n$ is a Borel function,
3. for each $s \leq t$ and $N \in \mathscr{N}^{\mathbb{P}}$, $\mathbb{1}_N \mathbb{1}_{\{\tau\leq s\}} \in \mathscr{V}$,

by the monotone class theorem [112, Theorems 1.2, 1.4], we conclude that $\mathscr{V}$ contains all $\sigma(\{\tau \leq s\} : s \leq t) \vee \mathscr{N}^{\mathbb{P}}$-measurable random variables.

Since $\mathscr{A}_t = \sigma(\{\tau \leq s\} : s \leq t) \vee \mathscr{N}^{\mathbb{P}}$ the proof is completed. □

2.1.1 A Key Lemma

The following lemma provides an essential tool to compute the conditional expectation of an integrable random variable given $\mathscr{A}_t$.

Lemma 2.3 *Let X be an integrable, $\mathscr{F}$-measurable r.v. Then,*[1]

$$\mathbb{E}[X|\mathscr{A}_t] = \mathbb{1}_{\{\tau>t\}}\frac{\mathbb{E}[X\mathbb{1}_{\{\tau>t\}}]}{\mathbb{P}(\tau > t)} + \mathbb{E}[X|\sigma(\tau)]\mathbb{1}_{\{\tau\leq t\}}\,. \tag{2.1}$$

[1] Here, and in the rest of the book, we write $\frac{1}{b}\mathbb{1}_{\{b>0\}}$ for the quantity equal to $\frac{1}{b}$ if $b > 0$ and equal to 0 if $b = 0$.

Proof Let $t^* := \inf\{s : F(s) = 1\}$. Then τ is bounded a.s. by t^* and $\mathbb{P}(\tau > t)$ is positive for $t^* > t$. Let $A \in \mathscr{A}_t$, then, by Lemma 2.2 there exists $\widehat{A} \in \mathscr{A}_0$ such that $\{\tau > t\} \cap A = \{\tau > t\} \cap \widehat{A}$, and we have

$$\mathbb{E}\left[X\mathbb{P}(\tau > t)\mathbb{1}_{\{\tau>t\}}\mathbb{1}_A\right] = \mathbb{E}\left[\mathbb{P}(\tau > t)\mathbb{1}_{\widehat{A}}\mathbb{E}\left[X\mathbb{1}_{\{\tau>t\}}\right]\right] = \mathbb{E}\left[\mathbb{E}[X\mathbb{1}_{\{\tau>t\}}]\mathbb{1}_{\{\tau>t\}}\mathbb{1}_A\right]$$

which leads to

$$\mathbb{E}\left[X\mathbb{1}_{\{\tau>t\}}|\mathscr{A}_t\right] = \mathbb{1}_{\{\tau>t\}}\frac{\mathbb{E}[X\mathbb{1}_{\{\tau>t\}}]}{\mathbb{P}(\tau > t)}.$$

Lemma 2.2 implies also that

$$\mathbb{E}\left[X\mathbb{1}_{\{\tau\leq t\}}|\mathscr{A}_t\right] = \mathbb{E}\left[\mathbb{E}[X|\sigma(\tau)]\mathbb{1}_{\{\tau\leq t\}}|\mathscr{A}_t\right] = \mathbb{E}[X|\sigma(\tau)]\mathbb{1}_{\{\tau\leq t\}}.$$

Thus, (2.1) is proven. □

2.1.2 Martingales and Predictable Representation Property

The Doob–Meyer decomposition of the increasing process A is derived in the next proposition. Particular attention is given to its martingale part M, which is a fundamental martingale enjoying the predictable representation property (PRP) in the filtration $\mathbb{A}$, as proven in Proposition 2.7.

Proposition 2.4 *The process M given by*

$$M_t := A_t - \int_0^{\tau\wedge t} \frac{dF(s)}{1 - F(s-)} \tag{2.2}$$

is an $\mathbb{A}$-martingale.

If F is absolutely continuous w.r.t. the Lebesgue measure with density f, the process

$$M_t := A_t - \int_0^{\tau\wedge t} \lambda(s)ds = A_t - \int_0^t \lambda(s)(1 - A_s)ds$$

is an $\mathbb{A}$-martingale, where $\lambda(s) = \dfrac{f(s)}{1 - F(s)}\mathbb{1}_{\{F(s)<1\}}$ is a deterministic non-negative function, called **the intensity rate of τ**.

Proof Note that

$$\int_0^{\tau\wedge t} \frac{dF(s)}{1 - F(s-)} = \int_0^t (1 - A_{s-})\frac{dF(s)}{1 - F(s-)}.$$

Let $t \le u$. Then, on the one hand, we obtain

$$\mathbb{E}[A_u - A_t|\mathscr{A}_t] = \mathbb{E}\left[\mathbb{1}_{\{t<\tau\le u\}}|\mathscr{A}_t\right] = \mathbb{1}_{\{t<\tau\}}\frac{F(u)-F(t)}{1-F(t)}, \tag{2.3}$$

where the second equality follows from Lemma 2.3 applied to $\mathbb{1}_{\{u\ge\tau\}}$. On the other hand, applying once again Lemma 2.3, we obtain

$$\mathbb{E}\left[\int_{(t\wedge\tau,u\wedge\tau]}\frac{dF(s)}{1-F(s-)}\,\Big|\,\mathscr{A}_t\right] = \mathbb{1}_{\{t<\tau\}}\frac{1}{\mathbb{P}(t<\tau)}\,\mathbb{E}\left[\int_{(t,u]}\mathbb{1}_{\{s\le\tau\}}\frac{dF(s)}{1-F(s-)}\right]$$
$$= \mathbb{1}_{\{t<\tau\}}\frac{1}{\mathbb{P}(t<\tau)}\int_{(t,u]}\mathbb{P}(s\le\tau)\,\frac{dF(s)}{1-F(s-)} = \mathbb{1}_{\{t<\tau\}}\frac{F(u)-F(t)}{1-F(t)}.$$

In view of (2.3), this proves the result. □

Example 2.5 Let N be an inhomogeneous Poisson process with deterministic intensity function λ and let τ be the first jump time of N. Then the default indicator process of τ is $A_t = N_{t\wedge\tau}$. Since $N_t - \int_0^t \lambda(s)ds$ is a martingale, it is also a martingale when stopped at time τ, i.e., $A_t - \int_0^{t\wedge\tau}\lambda(s)ds$, is a martingale. Therefore λ is the intensity rate of τ.

Corollary 2.6 *For any bounded Borel function $h : [0,\infty] \to \mathbb{R}$, the process M^h given by*

$$M_t^h := \mathbb{1}_{\{\tau\le t\}}h(\tau) - \int_0^{t\wedge\tau} h(u)\,\frac{dF(u)}{1-F(u-)}$$

satisfies $dM_t^h = h(t)dM_t$ where M is defined in (2.2) and is an $\mathbb{A}$-martingale.

Proof This is indeed true since

$$M_t^h = \int_0^t h(u)dA_u - \int_0^t (1-A_{u-})h(u)\,\frac{dF(u)}{1-F(u-)} = \int_0^t h(u)dM_u\,.$$

The martingale property follows from the fact that h is $\mathbb{A}$-predictable and bounded. □

We will generalize the above results in Sect. 2.2.3.

Proposition 2.7 *Any $\mathbb{A}$-local martingale can be written as a stochastic integral w.r.t. the martingale M defined in (2.2); in other terms, M has the PRP in the filtration $\mathbb{A}$.*

Proof Using the fact that $\mathscr{A}_\infty = \sigma(\tau)$, any bounded $\mathbb{A}$-martingale Y can be written as $Y_t = \mathbb{E}[h(\tau)|\mathscr{A}_t]$, where h is a bounded Borel function $h : [0,\infty] \to \mathbb{R}$. Then, by Lemma 2.3, we have

$$Y_t = \mathbb{E}[h(\tau)|\mathscr{A}_t] = \mathbb{1}_{\{\tau\le t\}}h(\tau) + \frac{\mathbb{1}_{\{t<\tau\}}}{1-F(t)}\left(\int_t^\infty h(u)\,dF(u) + h(\infty)(1-F(\infty))\right).$$

Recall that $\tau \leq t^*$ a.s. for $t^* := \inf\{t : F(t) = 1\}$. Therefore $Y_t = h(\tau)$ for $t \geq t^*$ and the previous formula implies that $Y_{t^*} = \lim_{t \nearrow t^*} Y_t$ since $\lim_{t \nearrow t^*} Y_t = h(\tau)$. Then, it is enough to show the integral representation of Y up to $t < t^*$. Note that, for $t < t^*$, one has $d(1 - F(t))^{-1} = ((1 - F(t))(1 - F(t-)))^{-1} dF(t)$. By integration by parts (see Proposition 1.16 (b)), for $t < t^*$, we firstly deduce that

$$d\frac{1 - A_t}{1 - F(t)} = \frac{1 - A_{t-}}{(1 - F(t))(1 - F(t-))} dF(t) - \frac{dA_t}{1 - F(t)}$$

and secondly that

$$d\left(\frac{1 - A_t}{1 - F(t)} \int_t^\infty h(u)\, dF(u)\right) = -\frac{1 - A_{t-}}{1 - F(t-)} h(t) dF(t)$$
$$+ \left(\int_t^\infty h(u)\, dF(u)\right) \frac{1 - A_{t-}}{(1 - F(t))(1 - F(t-))} dF(t) - \left(\int_t^\infty h(u)\, dF(u)\right) \frac{dA_t}{1 - F(t)}.$$

Finally, denoting

$$K(t) := \frac{1}{1 - F(t)} 1\!\!1_{\{F(t)<1\}} \left(\int_t^\infty h(u)\, dF(u) + h(\infty)(1 - F(\infty))\right),$$

we conclude that $dY_t = (h(t) - K(t)) dM_t$ for each $t \geq 0$. It is then standard, by localizing argument that the PRP is valid for any local martingale. □

2.2 Compensator of the Default Process in a General Setting

In this section, a probability space $(\Omega, \mathscr{F}, \mathbb{P})$, endowed with a filtration $\mathbb{F} := (\mathscr{F}_t, t \geq 0)$ satisfying the usual conditions of $\mathbb{P}$-completeness, right-continuity and such that $\mathscr{F}_\infty \subset \mathscr{F}$, is given. We define two filtrations $\mathbb{G}^0 := (\mathscr{G}_t^0)_{t\geq 0}$ and $\mathbb{G} := (\mathscr{G}_t)_{t \geq 0}$ by

$$\mathscr{G}_t^0 := \mathscr{F}_t \vee \mathscr{A}_t \quad \text{and} \quad \mathscr{G}_t := \cap_{\varepsilon > 0} \mathscr{G}_{t+\varepsilon}^0, \tag{2.4}$$

or, equivalently $\mathbb{G}^0 := \mathbb{F} \vee \mathbb{A}$ and $\mathbb{G} := \mathbb{F} \triangledown \mathbb{A}$ (see notation on p. 1). In other terms, $\mathbb{G}$ is the progressively enlarged filtration which is the smallest right-continuous filtration containing $\mathbb{F}$ and making τ a stopping time.

From Proposition 1.46, the Doob–Meyer decomposition of the $\mathbb{F}$-Azéma supermartingale $Z := {}^{o,\mathbb{F}}(1 - A)$, or equivalently $Z_t = \mathbb{P}(\tau > t | \mathscr{F}_t)$, is $Z = n - A^p$ where n is an $\mathbb{F}$-martingale and A^p is the $\mathbb{F}$-dual predictable projection of the increasing process $A := 1\!\!1_{[\![\tau, \infty[\![}$.

2.2.1 A Key Lemma

Lemma 2.9 provides an essential tool to compute the conditional expectation of an integrable random variable given $\mathscr{G}_t$ in terms of the conditional expectations given $\mathscr{F}_t$. Before we proceed to that key lemma we establish the following proposition which gives the equivalent form of $\mathscr{G}_t^0$-measurable r.v.'s.

Proposition 2.8 *For a fixed t, a random variable Y is $\mathscr{G}_t^0$-measurable if and only if it is of the form*

$$Y = y 1\!\!1_{\{t<\tau\}} + \widehat{y}(\tau) 1\!\!1_{\{\tau \le t\}} \tag{2.5}$$

for an $\mathscr{F}_t$-measurable random variable y and an $\mathscr{F}_t \otimes \mathscr{B}(\overline{\mathbb{R}}^+)$-measurable mapping $(\omega, u) \to \widehat{y}(\omega, u)$.

Proof Let $\mathscr{V}$ be the vector space of random variables defined as follows

$$\mathscr{V} := \{Y : Y = y 1\!\!1_{\{t<\tau\}} + \widehat{y}(\tau) 1\!\!1_{\{\tau \le t\}} \text{ for } \mathscr{F}_t\text{ -measurable r.v. } y \\ \text{and } \mathscr{F}_t \otimes \mathscr{B}(\overline{\mathbb{R}}^+)\text{-measurable mapping } \widehat{y}\}.$$

First note that any r.v. $Y \in \mathscr{V}$ is $\mathscr{G}_t^0$-measurable since, for any $B \in \mathscr{B}(\mathbb{R})$, we have

$$Y^{-1}(B) = \{\omega : y \in B,\ t < \tau(\omega)\} \cup \{\omega : \widehat{y}(\tau(\omega) \wedge t) \in B, \tau(\omega) \le t\} \in \mathscr{G}_t^0$$

as $\tau \wedge t$ is a $\mathscr{G}_t^0$-measurable random variable. On the other hand, $\mathscr{V}$ satisfies that (i) $1 \in \mathscr{V}$; (ii) if $Y^n := y^n 1\!\!1_{\{t<\tau\}} + \widehat{y}^n(\tau) 1\!\!1_{\{\tau \le t\}}$ where y^n and $\widehat{y}^n$ are respectively $\mathscr{F}_t$-measurable r.v. and $\mathscr{F}_t \otimes \mathscr{B}(\overline{\mathbb{R}}^+)$-measurable mapping, $Y^n \nearrow Y$ and Y is finite, then $Y = y 1\!\!1_{\{t<\tau\}} + \widehat{y}(\tau) 1\!\!1_{\{\tau \le t\}}$, where $y := \limsup y^n$ and $\widehat{y} := \limsup \widehat{y}^n$ are measurable accordingly; and (iii) for each $s \le t$ and $F \in \mathscr{F}_t$, $1\!\!1_F 1\!\!1_{\{\tau \le s\}} \in \mathscr{V}$. By the monotone class theorem, we conclude that $\mathscr{V}$ contains all $\sigma(\{\tau \le s\} : s \le t) \vee \mathscr{F}_t$-measurable random variables which completes the proof. □

Lemma 2.9 *Let X be an $\mathscr{F}$-measurable integrable r.v. Then, for any $t \ge 0$,*

$$\mathbb{E}\left[X 1\!\!1_{\{\tau>t\}} | \mathscr{G}_t\right] = 1\!\!1_{\{\tau>t\}} \frac{1}{Z_t} \mathbb{E}\left[X 1\!\!1_{\{\tau>t\}} | \mathscr{F}_t\right].$$

Proof The right-hand side of the above identity is well-defined since for each $t \ge 0$ one has $0 = \mathbb{E}[1\!\!1_{\{Z_t=0\}} Z_t] = \mathbb{E}[1\!\!1_{\{Z_t=0\}} 1\!\!1_{\{\tau>t\}}]$ which implies $\{\tau > t\} \subset \{Z_t > 0\}$. By Proposition 2.8, for any $G \in \mathscr{G}_t^0$ there exists $F \in \mathscr{F}_t$ s.t. $G \cap \{\tau > t\} = F \cap \{\tau > t\}$. Hence:

$$\begin{aligned}
\mathbb{E}\left[X 1\!\!1_{\{\tau>t\}\cap G} Z_t\right] &= \mathbb{E}\left[X 1\!\!1_{\{\tau>t\}\cap F} Z_t\right] = \mathbb{E}\left[\mathbb{E}\left[X 1\!\!1_{\{\tau>t\}} | \mathscr{F}_t\right] 1\!\!1_F Z_t\right] \\
&= \mathbb{E}\left[\mathbb{E}\left[X 1\!\!1_{\{\tau>t\}} | \mathscr{F}_t\right] 1\!\!1_F \mathbb{E}\left[1\!\!1_{\{\tau>t\}} | \mathscr{F}_t\right]\right] = \mathbb{E}\left[\mathbb{E}\left[X 1\!\!1_{\{\tau>t\}} | \mathscr{F}_t\right] 1\!\!1_{F\cap\{\tau>t\}}\right] \\
&= \mathbb{E}\left[\mathbb{E}\left[X 1\!\!1_{\{\tau>t\}} | \mathscr{F}_t\right] 1\!\!1_{G\cap\{\tau>t\}}\right]
\end{aligned}$$

which shows that

$$\mathbb{E}\left[X\mathbb{1}_{\{\tau>t\}}|\mathcal{G}_t^0\right] = \mathbb{1}_{\{\tau>t\}}\frac{1}{Z_t}\mathbb{E}[X\mathbb{1}_{\{\tau>t\}}|\mathcal{F}_t].$$

By Theorem 1.10 the right-hand side in the above equality has a right-continuous modification (as a process) and, by Theorem 1.11, the following equalities hold:

$$\mathbb{E}\left[X\mathbb{1}_{\{\tau>t\}}|\mathcal{G}_t\right] = \mathbb{E}\left[X\mathbb{1}_{\{\tau>t\}}|\mathcal{G}_{t+}^0\right] = \lim_{u\downarrow t}\mathbb{E}\left[X\mathbb{1}_{\{\tau>t\}}|\mathcal{G}_u^0\right] = \mathbb{E}\left[X\mathbb{1}_{\{\tau>t\}}|\mathcal{G}_t^0\right].$$

That completes the proof. □

Combining Lemma 2.9 and the definition of dual projections provides us with the following result.

Corollary 2.10 *(a) Let h be a bounded $\mathbb{F}$-optional process. Then*

$$\mathbb{E}[h_\tau|\mathcal{G}_t] = h_\tau\mathbb{1}_{\{\tau\le t\}} + \mathbb{1}_{\{\tau>t\}}\frac{1}{Z_t}\mathbb{E}\left[\int_t^\infty h_u dA_u^o + h_\infty Z_\infty\Big|\mathcal{F}_t\right]. \tag{2.6}$$

(b) Let h be a bounded $\mathbb{F}$-predictable process. Then

$$\mathbb{E}[h_\tau|\mathcal{G}_t] = h_\tau\mathbb{1}_{\{\tau\le t\}} + \mathbb{1}_{\{\tau>t\}}\frac{1}{Z_t}\mathbb{E}\left[\int_t^\infty h_u dA_u^p + h_\infty Z_\infty\Big|\mathcal{F}_t\right]. \tag{2.7}$$

Proof By Lemma 2.9 one has

$$\mathbb{E}[h_\tau|\mathcal{G}_t] = h_\tau\mathbb{1}_{\{\tau\le t\}} + \mathbb{1}_{\{\tau>t\}}\frac{1}{Z_t}\mathbb{E}[h_\tau\mathbb{1}_{\{t<\tau\le\infty\}}|\mathcal{F}_t].$$

We note that $h_\tau\mathbb{1}_{\{t<\tau\}} = \int_t^\infty h_u dA_u + h_\infty\mathbb{1}_{\{\tau=\infty\}}$. Then the assertion (a) follows by Definition 1.39 (a) of the $\mathbb{F}$-dual optional projection of τ.

The proof of assertion (b) follows by an analogous argument, using the $\mathbb{F}$-dual predictable projection of τ (see Definition 1.39 (b)). □

2.2.2 $\mathbb{G}$-Measurability Versus $\mathbb{F}$-Measurability

Proposition 2.11 *(a) For a $\mathbb{G}$-optional process Y, the $\mathbb{F}$-optional process y defined as $y := {}^{o,\mathbb{F}}(Y\mathbb{1}_{[\![0,\tau[\![})\mathbb{1}_{\{Z>0\}}\frac{1}{Z}$ satisfies*

$$\mathbb{1}_{[\![0,\tau[\![}Y = \mathbb{1}_{[\![0,\tau[\![}\, y\,. \tag{2.8}$$

The process y is called the $(\mathbb{F},\tau)$-optional reduction of Y.

(b) A process Y is $\mathbb{G}$-predictable if and only if it is of the form

$$Y = \mathbb{1}_{[\![0,\tau]\!]} y + \mathbb{1}_{]\!]\tau,\infty[\![} \, \widehat{y}(\tau) \tag{2.9}$$

where y is $\mathbb{F}$-predictable and $(\omega, t, u) \mapsto \widehat{y}_t(\omega, u)$ is a $\mathscr{P}(\mathbb{F}) \otimes \mathscr{B}(\overline{\mathbb{R}}^+)$-measurable function. Moreover one can choose $y := {}^{p,\mathbb{F}}(Y \mathbb{1}_{[\![0,\tau]\!]}) \mathbb{1}_{\{Z_- > 0\}} \frac{1}{Z_-}$ which is called the $(\mathbb{F}, \tau)$-predictable reduction of Y.

Proof (a) Stochastic processes of the form $f(t)X(\omega)$ where f is a Borel function and X is an $\mathscr{F}$-measurable r.v. generate all stochastic processes by the monotone class theorem (see Definition 1.1 where stochastic processes are required to be measurable). Hence the processes of the form $f(t)X_t(\omega)$ where f is a Borel function and X is a càdlàg $\mathbb{G}$-martingale generate all $\mathbb{G}$-optional processes. It is therefore enough to establish (2.8) for any càdlàg $\mathbb{G}$-martingale X. Lemma 2.9 implies that for any $t \geq 0$ and integrable r.v. X_∞,

$$X_t \mathbb{1}_{\{t<\tau\}} = \mathbb{E}[X_\infty | \mathscr{G}_t] \mathbb{1}_{\{t<\tau\}} = \mathbb{1}_{\{\tau>t\}} \frac{\mathbb{E}\left[X_\infty \mathbb{1}_{\{\tau>t\}} | \mathscr{F}_t\right]}{Z_t}.$$

Note that, by Theorem 1.10, the process $\mathbb{E}\left[X_\infty \mathbb{1}_{\{\tau>t\}} | \mathscr{F}_t\right] Z_t^{-1} \mathbb{1}_{\{Z_t>0\}}$ has a càdlàg modification x. Hence, since x is $\mathbb{F}$-adapted, x is also an $\mathbb{F}$-optional process satisfying $\mathbb{1}_{[\![0,\tau[\![} X = \mathbb{1}_{[\![0,\tau[\![} x$. Therefore, for any $\mathbb{G}$-optional process Y there exists an $\mathbb{F}$-optional process $\widetilde{y}$ such that $\mathbb{1}_{[\![0,\tau[\![} Y = \mathbb{1}_{[\![0,\tau[\![} \widetilde{y}$ and taking $\mathbb{F}$-optional projections:

$${}^{o}(\mathbb{1}_{[\![0,\tau[\![} Y) = {}^{o}\, (\mathbb{1}_{[\![0,\tau[\![} \, \widetilde{y}) = {}^{o}(\mathbb{1}_{[\![0,\tau[\![}) \, \widetilde{y} = \; Z \, \widetilde{y}.$$

Hence $\widetilde{y}$ is only uniquely determined on $\{Z > 0\}$ and satisfies

$$\mathbb{1}_{\{Z>0\}} \widetilde{y} = \frac{{}^{o}(\mathbb{1}_{[\![0,\tau[\![} Y)}{Z} \mathbb{1}_{\{Z>0\}}.$$

The $(\mathbb{F}, \tau)$-reduction y consists of choosing the process satisfying $y \mathbb{1}_{\{Z=0\}} = 0$.
(b) A process given by the right-hand side of (2.9) where $\widehat{y}_t(\omega, u) = \widetilde{y}_t(\omega) f(u)$, with a Borel function $f : \overline{\mathbb{R}}^+ \to \mathbb{R}$ and an $\mathbb{F}$-predictable process $\widetilde{y}$, is clearly $\mathbb{G}$-predictable. The statement for a general $\widehat{y}$ follows by the usual argument based on the monotone class theorem.

To show that any $\mathbb{G}$-predictable process is of the given form, also by the monotone class theorem, it is enough to consider the elements from the generator of $\mathbb{G}$-predictable processes. Since, by Theorem 3.21 in [112], $\mathbb{G}$-predictable processes are generated by the processes of the form $Y_t(\omega) = f(s \wedge \tau(\omega)) F_s(\omega) \mathbb{1}_{\{t \geq s\}}$, where s is a fixed real number, f is a Borel function and F_s is an $\mathscr{F}_s$-measurable r.v., we conclude that (2.9) holds.

The second part of the statement about the form of $(\mathbb{F}, \tau)$-predictable reduction of Y follows by an argument analogous to the one used in the proof of (b). □

Corollary 2.12 *For any $\mathbb{G}$-stopping time S there exists an $\mathbb{F}$-stopping time T such that $S \wedge \tau = T \wedge \tau$. For any $\mathbb{G}$-predictable stopping time S there exists an $\mathbb{F}$-predictable stopping time T such that $S \wedge \tau = T \wedge \tau$.*

Proof In the first case it is enough to apply Proposition 2.11 (b) to the process $\mathbb{1}_{[\![0,S]\!]}$, and in the second case to $\mathbb{1}_{[\![0,S[\![}$. □

Remark 2.13 A characterization result analogous to (2.9) for optional processes does *not* hold in general. We refer to Barlow's counterexample presented in Example 5.13 and Corollary 5.12 which concerns the case of an honest time.

Lemma 2.14 *The process Z is positive on the stochastic interval $[\![0,\tau[\![$ and the processes Z_- and $\widetilde{Z}$ are positive on the stochastic interval $[\![0,\tau]\!]$.*

Proof We shall use the optional section theorem to prove that the $\mathbb{G}$-optional set $\{Z = 0\} \cap [\![0,\tau[\![$ is evanescent. Corollary 2.12 implies that for any $\mathbb{G}$-stopping time S there exists an $\mathbb{F}$-stopping time T such that $S \wedge \tau = T \wedge \tau$. For any $\mathbb{F}$-stopping time T, it holds $0 = \mathbb{E}[\mathbb{1}_{\{Z_T=0\}} Z_T] = \mathbb{E}[\mathbb{1}_{\{Z_T=0\}} \mathbb{1}_{\{\tau>T\}}]$. Therefore the supermartingale Z is positive on the interval $[\![0,\tau[\![$.

Using a similar argument based on the predictable section theorem and the fact that $0 = \mathbb{E}[\mathbb{1}_{\{Z_{T-}=0\}} Z_{T-}] = \mathbb{E}[\mathbb{1}_{\{Z_{T-}=0\}} \mathbb{1}_{\{\tau \geq T\}}]$ for any $\mathbb{F}$-predictable stopping time T, we conclude that the $\mathbb{G}$-predictable set $\{Z_- = 0\} \cap [\![0,\tau]\!]$ is evanescent. Therefore the left-limit Z_- is positive on $[\![0,\tau]\!]$. The proof for $\widetilde{Z}$ is analogous. □

2.2.3 Martingales

The generalisation of Proposition 2.4 to the case of a non-trivial reference filtration $\mathbb{F}$ is given below.

Proposition 2.15 *The process M given by*

$$M_t := A_t - \int_0^{t\wedge\tau} \frac{1}{Z_{u-}} dA^p_u \tag{2.10}$$

is a u.i. $\mathbb{G}$-martingale. In other words, the process $(1 - A_-)\frac{1}{Z_-} \cdot A^p$ is the $\mathbb{G}$-compensator of τ, and its $(\mathbb{F},\tau)$-predictable reduction equals $\frac{1}{Z_-}\mathbb{1}_{\{Z_->0\}} \cdot A^p$.

Proof Note that, by Lemma 2.14, the process $(1 - A_-)\frac{1}{Z_-}$ is well-defined. The assertion is a consequence of Proposition 1.36 (a) as soon as we show the desired form of the $\mathbb{G}$-dual predictable projection of A (or $\mathbb{G}$-compensator of A, since A is $\mathbb{G}$-adapted). It is equivalent to prove that, for any $\mathbb{G}$-predictable bounded process H, one has

$$\mathbb{E}\left[\int_{[0,\infty)} H_s dA_s\right] = \mathbb{E}\left[\int_{[0,\infty)} H_s(1 - A_{s-})\frac{dA^p_s}{Z_{s-}}\right].$$

Let h be the $(\mathbb{F},\tau)$-predictable reduction of H (see Proposition 2.11 (b)). By the definition of the dual predictable projection,

$$\mathbb{E}\left[\int_{[0,\infty)} H_s dA_s\right] = \mathbb{E}\left[\int_{[0,\infty)} h_s \mathbb{1}_{\{Z_{s-}>0\}} dA_s\right] = \mathbb{E}\left[\int_{[0,\infty)} h_s \mathbb{1}_{\{Z_{s-}>0\}} dA^p_s\right].$$

Identities (1.9) and (1.18) imply that the last term equals

$$\mathbb{E}\left[\int_{[0,\infty)} \frac{h_s}{Z_{s-}} {}^p(1-A_-)_s \mathbb{1}_{\{Z_{s-}>0\}} dA^p_s\right] = \mathbb{E}\left[\int_{[0,\infty)} \frac{h_s}{Z_{s-}}(1-A_{s-}) dA^p_s\right]$$

and the result follows by recalling that h is the $(\mathbb{F},\tau)$-predictable reduction of H. □

Similarly to Corollary 2.6, the following result is in force.

Corollary 2.16 *For a bounded $\mathbb{G}$-predictable process H, the process M^h given by*

$$M^h_t := H_\tau A_t - \int_0^{t\wedge\tau} \frac{H_u}{Z_{u-}} dA^p_u$$

satisfies $dM^h_t = H_t\, dM_t$ where M is defined in (2.10) and is a u.i. $\mathbb{G}$-martingale.

In the next proposition the distribution of the $\mathbb{G}$-compensator of τ sampled at time τ is studied. We also refer to the connected result given in Sect. 3.3.2.

Proposition 2.17 *Assume that τ is a finite random time such that A^p is continuous. Then the random variable $\Lambda_\tau := \int_0^\tau \frac{1}{Z_{u-}} dA^p_u$ has a unit exponential law.*

Proof Denote the $\mathbb{G}$-compensator of A by Λ, i.e., $\Lambda_t = \int_0^{\tau\wedge t} \frac{1}{Z_{u-}} dA^p_u$ (see Proposition 2.15) and note that the definition of r.v. Λ_τ is consistent with this notation. Let φ be a bounded Borel function, $\Phi(t) = \int_0^t \varphi(s)ds$ and consider the martingale

$$\int_0^t \varphi(\Lambda_s) dM_s = \varphi(\Lambda_\tau)\mathbb{1}_{\{\tau\le t\}} - \int_0^t \varphi(\Lambda_s) d\Lambda_s = \varphi(\Lambda_\tau)\mathbb{1}_{\{\tau\le t\}} - \Phi(\Lambda_t).$$

Then, with $t\to\infty$, using the fact that $\Lambda_\infty = \Lambda_\tau$, one has $\mathbb{E}[\varphi(\Lambda_\tau)] = \mathbb{E}[\Phi(\Lambda_\tau)]$. Taking $\varphi(t) = -ae^{-at}$, one gets $\mathbb{E}[\exp(-a\Lambda_\tau)] = (1+a)^{-1}$ which characterizes the unit exponential law. □

Proposition 2.18 *Define the process Υ by $\Upsilon := (1-A)Z^{-1}$.*
(a) Let X be a non-negative $\mathbb{F}$-supermartingale. Then, $X\Upsilon$ is a $\mathbb{G}$-supermartingale.
(b) Assume that Z is positive and X is an $\mathbb{F}$-martingale. Then $X\Upsilon$ is a $\mathbb{G}$-martingale.
(c) If the process Z is positive, decreasing and continuous, then the $\mathbb{G}$-compensator of τ equals $\Lambda^\tau := (\Lambda_{t\wedge\tau}, t\ge 0)$ where $\Lambda := -\ln Z$. In other words, the $\mathbb{G}$-martingale M, introduced in (2.10), is $M = A - \Lambda^\tau$. Moreover $\Upsilon = \Upsilon_0\mathcal{E}(-M)$.

Proof Note that Υ is well-defined since the random sets $\{Z = 0\}$ and $[\![0, \tau[\![$ are disjoint as shown in Lemma 2.14.
(a) For $s \le t$, one has

$$\mathbb{E}\left[\Upsilon_t X_t | \mathcal{G}_s\right] = \mathbb{E}\left[\mathbb{1}_{\{\tau>t\}} \frac{1}{Z_t} X_t \Big| \mathcal{G}_s\right] = \mathbb{1}_{\{\tau>s\}} \frac{1}{Z_s} \mathbb{E}\left[\mathbb{1}_{\{\tau>t\}} \frac{1}{Z_t} X_t \Big| \mathcal{F}_s\right]$$

$$= \mathbb{1}_{\{\tau>s\}} \frac{1}{Z_s} \mathbb{E}\left[\mathbb{E}\left[\mathbb{1}_{\{\tau>t\}} | \mathcal{F}_t\right] \mathbb{1}_{\{Z_t>0\}} \frac{1}{Z_t} X_t \Big| \mathcal{F}_s\right] = \Upsilon_s \mathbb{E}\left[X_t \mathbb{1}_{\{Z_t>0\}} | \mathcal{F}_s\right] \le \Upsilon_s X_s \,,$$

where the second equality holds by Lemma 2.9.
(b) This is a consequence of similar computations as (a).
(c) Under the hypotheses on Z, Proposition 2.15 implies that the $\mathbb{G}$-compensator of A is $-(1 - A_-)\frac{1}{Z} \cdot Z = \Lambda^\tau$, hence the first assertion holds true. It remains to show that $d\Upsilon_t = -\Upsilon_{t-} dM_t$ which follows from integration by parts and $A^p = 1 - Z$. □

Remark 2.19 (a) We illustrate the difference between (a) and (b) in the above proposition with a trivial example. Let τ be an $\mathbb{F}$-stopping time, then $Z = \mathbb{1}_{[\![0,\tau[\![}$ and it vanishes at time τ. The process $\Upsilon = (1 - A)$ is indeed a $\mathbb{G}$-supermartingale but is not a $\mathbb{G}$-martingale.
(b) Assertion (b) in the above proposition seems to be related to a change of probability. It is important to note that here, one changes the filtration, not the probability measure. Moreover, setting $\frac{d\mathbb{Q}}{d\mathbb{P}}|_{\mathcal{G}_t} = \Upsilon_t$ does not define a probability $\mathbb{Q}$ *equivalent* to $\mathbb{P}$, since the non-negative martingale Υ vanishes. The probability $\mathbb{Q}$ would be only absolutely continuous w.r.t. $\mathbb{P}$. See Collin-Dufresne et al. [64].
(c) Assume that Z is continuous and positive. We recall that the Doob–Meyer decomposition of Z is $Z = n - A^p$. By integration by parts one obtains

$$d\Upsilon_t = -(1 - A_{t-})\frac{1}{Z_t^2}\left(dn_t - dA_t^p\right) + \frac{1 - A_{t-}}{Z_t^3} d\langle n\rangle_t^{\mathbb{F}} - \frac{1}{Z_t} dA_t$$

and it follows that

$$d\Upsilon_t + \frac{1}{Z_t} dM_t = -(1 - A_{t-})\frac{1}{Z_t^2}\left(dn_t - \frac{1}{Z_t} d\langle n\rangle_t^{\mathbb{F}}\right).$$

Due to the $\mathbb{G}$-martingale property of Υ and M, the quantity $(1 - A_-)\frac{1}{Z^2} \cdot (n - \frac{1}{Z} \cdot \langle n\rangle^{\mathbb{F}})$ must be a $\mathbb{G}$-local martingale (stopped at τ). We shall see, in Chap. 5, that the martingale property of $n - \frac{1}{Z} \cdot \langle n\rangle^{\mathbb{F}}$ stopped at τ is a consequence of the general Jeulin's formula (5.1).

Definition 2.20 If there exists a $\mathbb{G}$-predictable (resp. $\mathbb{F}$-predictable) process $\lambda^{\mathbb{G}}$ (resp. $\lambda^{\mathbb{F}}$) such that

$$A_t - \int_0^t \lambda_s^{\mathbb{G}} ds = A_t - \int_0^{\tau\wedge t} \lambda_s^{\mathbb{F}} ds$$

is a $\mathbb{G}$-martingale, then $\lambda^{\mathbb{G}}$ (resp. $\lambda^{\mathbb{F}}$) is called the **$\mathbb{G}$-intensity rate of** τ (resp. the **$\mathbb{F}$-intensity rate of** τ).

Remark 2.21 Assume that A^p is absolutely continuous w.r.t. the Lebesgue measure, i.e., there exists a measurable process $\widetilde{a}$ such that $A^p = \int_0^{\cdot} \widetilde{a}_s ds$. Note that ${}^p\widetilde{a}$ exists since it is non-negative. Since $(A^p)^p = A^p$, identity (1.9) implies that $A^p = \int_0^{\cdot} a_s ds$ where $a = {}^p\widetilde{a}$ is $\mathbb{F}$-predictable. It was proven in Proposition 2.15 that the process

$$A_t - \int_0^{t\wedge\tau} \lambda_u du = A_t - \int_0^t (1 - A_{u-})\lambda_u du$$

where $\lambda_u = \frac{a_u}{Z_{u-}} 1\!\!1_{\{Z_{u-}>0\}}$ is a $\mathbb{G}$-martingale. Therefore λ is the $\mathbb{F}$-intensity rate of τ and $\lambda 1\!\!1_{[\![0,\tau]\!]}$ is the $\mathbb{G}$-intensity rate of τ.

The Ethier–Kurtz Criterion establishes that, if there exists $K < \infty$ such that for any $s < t$, $\mathbb{E}[A_t - A_s|\mathscr{G}_s] \leq K(t-s)$, then A^p is absolutely continuous (see [93, 126]).

Lemma 2.22 *The $\mathbb{F}$-intensity rate process λ, if it exists, satisfies*

$$\lambda_t = \lim_{h\to 0} \frac{1}{h} \frac{\mathbb{P}(t < \tau < t+h|\mathscr{F}_t)}{\mathbb{P}(t < \tau|\mathscr{F}_t)} .$$

Proof The martingale property of M implies that

$$\mathbb{E}[1\!\!1_{\{t<\tau\leq t+h\}}|\mathscr{G}_t] = \int_t^{t+h} \mathbb{E}[(1 - A_s)\lambda_s|\mathscr{G}_t]ds .$$

It follows that, on $\{t < \tau\}$

$$\lambda_t = \lim_{h\to 0} \frac{1}{h}\mathbb{P}(t < \tau < t+h|\mathscr{G}_t) = \lim_{h\to 0} \frac{1}{h} \frac{\mathbb{P}(t < \tau < t+h|\mathscr{F}_t)}{\mathbb{P}(t < \tau|\mathscr{F}_t)} .$$

□

Proposition 2.23 *Suppose that the filtration $\widehat{\mathbb{F}}$ satisfies $\widehat{\mathbb{F}} \subset \mathbb{F}$. Denote by $\widehat{\mathbb{G}}$ the filtration $\widehat{\mathbb{G}} = \widehat{\mathbb{F}} \vee \mathbb{A}$, by $\widehat{Z}$ the $\widehat{\mathbb{F}}$-Azéma supermartingale of τ and by $A^{p,\widehat{\mathbb{F}}}$ the $\widehat{\mathbb{F}}$-dual predictable projection of A. Suppose that the $\mathbb{F}$-dual predictable projection of A is absolutely continuous w.r.t. the Lebesgue measure, i.e., $A^p = \int_0^{\cdot} a_s ds$ for an $\mathbb{F}$-predictable process a. Then, the $\widehat{\mathbb{F}}$-intensity rate of τ equals*

$${}^{p,\widehat{\mathbb{F}}}a_s \frac{1}{\widehat{Z}_{s-}} 1\!\!1_{\{\widehat{Z}_{s-}>0\}}.$$

Proof By virtue of the Proposition 2.15, Definition 2.20 and Remark 2.21, it is enough to check that $A^{p,\widehat{\mathbb{F}}} = \int_0^{\cdot} {}^{p,\widehat{\mathbb{F}}}a_s ds$ which follows from (1.9) and $A^{p,\widehat{\mathbb{F}}} = (A^p)^{p,\widehat{\mathbb{F}}}$. □

2.3 Cox Model and Extensions

This section is devoted to a particular construction of random times. It is a fundamental construction of a default time in finance. In a credit risk setting, the random variable τ represents the time when a default occurs. In the literature, models for default times are often based on a threshold: the default occurs when some driving process X passes a given barrier. Based on this observation, we consider a random time in a general threshold model. Let X be a stochastic process and Θ be a barrier which shall be made precise later. Define a random time as the first passage time at the level Θ:

$$\tau := \inf\{t : X_t \geq \Theta\}.$$

In classical structural models, a reference filtration $\mathbb{F}$ is given, the process X is $\mathbb{F}$-adapted and represents the value of a firm and the barrier Θ is a constant. So, τ is an $\mathbb{F}$-stopping time. If τ is an $\mathbb{F}$-predictable stopping time (e.g. if $\mathbb{F}$ is a Brownian filtration), the $\mathbb{F}$-compensator of $A := \mathbb{1}_{[\![\tau,\infty[\![}$ is A. The goal is then to compute the conditional law of the default $\mathbb{P}(\tau > \theta|\mathcal{F}_t)$, for $\theta > t$.

In a reduced form approach (say, if τ is not the first time where an observable process passes a constant barrier) there are two sources of information: information arriving from market prices modelled via a reference filtration $\mathbb{F}$ and the information about the default time, i.e., the knowledge whether the default occurred or not, modelled via filtration $\mathbb{A}$.

At the intuitive level, $\mathbb{F}$ is generated by prices of some assets, or by other economic factors (e.g. interest rates). The case where $\mathbb{F}$ is the trivial filtration is studied in Sect. 2.1. Though in typical examples $\mathbb{F}$ is chosen to be the Brownian filtration, most theoretical results do not rely on such a specification of the filtration $\mathbb{F}$.

2.3.1 *Construction of a Cox Model with a Given Intensity*

Let $(\Omega, \mathcal{F}, \mathbb{F}, \mathbb{P})$ be a filtered probability space. Assume that λ is a non-negative $\mathbb{F}$-adapted process and that there exists a random variable Θ, independent of $\mathcal{F}_\infty$, with a unit exponential law. In the **Cox model**, the default time τ is defined as the first time when the increasing process $\Lambda_t := \int_0^t \lambda_s\, ds$ crosses the random level Θ, i.e.,

$$\tau := \inf\{t \geq 0 : \Lambda_t \geq \Theta\}.$$

In particular, using the increasing property of Λ, one gets $\{\tau > s\} = \{\Lambda_s < \Theta\}$. We assume that $\Lambda_t < \infty$, for all t and $\Lambda_\infty = \infty$, in particular τ is a finite random time.

Remark 2.24 (a) If a probability space $(\Omega, \mathcal{F}, \mathbb{P})$ is given, in order to construct a r.v. Θ independent of $\mathcal{F}$, one may need to enlarge the probability space as follows. Let $(\widehat{\Omega}, \widehat{\mathcal{F}}, \widehat{\mathbb{P}})$ be an auxiliary probability space with a r.v. Θ with exponential law. We

introduce the product probability space $(\widetilde{\Omega}, \widetilde{\mathscr{F}}, \widetilde{\mathbb{P}}) = (\Omega \times \widehat{\Omega}, \mathscr{F} \otimes \widehat{\mathscr{F}}, \mathbb{P} \otimes \widehat{\mathbb{P}})$.
(b) Some authors define the default time as

$$\tau = \inf\{t \geq 0 \,:\, X_t \geq \Theta\} = \inf\left\{t \geq 0 \,:\, \sup_{s \leq t} X_s \geq \Theta\right\}$$

where X is a given $\mathbb{F}$-semimartingale.
(c) One can define the time of default as $\tau = \inf\{t \,:\, \Lambda_t \geq \Sigma\}$ where Σ is a non-negative r.v. independent of $\mathscr{F}_\infty$. Assume that the cumulative distribution function of Σ, denoted by Φ, is continuous and increasing. This model then reduces to the previous one: the r.v. $\Phi(\Sigma)$ has a uniform distribution and

$$\tau = \inf\{t \,:\, \Phi(\Lambda_t) \geq \Phi(\Sigma)\} = \inf\left\{t \,:\, \Psi^{-1} \circ \Phi(\Lambda_t) \geq \Theta\right\}$$

where Ψ is the cumulative distribution function of the unit exponential law.

2.3.2 Conditional Expectations and Immersion

Lemma 2.25 *The conditional distribution of τ given $\mathscr{F}_t$ for $t \in [0, \infty]$ equals*

$$\mathbb{P}(\tau > \theta | \mathscr{F}_t) = \mathbb{E}\left[\exp\left(-\Lambda_\theta\right) | \mathscr{F}_t\right] \quad \text{for} \quad \theta \in \mathbb{R}^+.$$

Equivalently, the conditional density function of τ given $\mathscr{F}_t$ for $t \in [0, \infty]$ equals

$$\mathbb{P}(\tau \in d\theta | \mathscr{F}_t) = \mathbb{E}\left[\lambda_\theta \exp(-\Lambda_\theta) | \mathscr{F}_t\right] d\theta \quad \text{for} \quad \theta \in \mathbb{R}^+.$$

In particular, the Azéma supermartingale $Z = \exp(-\Lambda)$ is decreasing, positive and continuous and satisfies $Z_t = \mathbb{P}(\tau > t | \mathscr{F}_\infty)$ and the dual predictable projection of τ equals $A^p = 1 - \exp(-\Lambda)$.

Proof The proof follows from the equality $\{\tau > \theta\} = \{\Lambda_\theta < \Theta\}$ for $\theta \in \mathbb{R}^+$ and the independence of Θ and $\mathscr{F}_\infty$. The $\mathscr{F}_t$-measurability of Λ_θ for $\theta \leq t$, implies

$$\mathbb{P}(\tau > \theta | \mathscr{F}_t) = \mathbb{P}\left(\Lambda_\theta < \Theta \mid \mathscr{F}_t\right) = \exp\left(-\Lambda_\theta\right).$$

The $\mathscr{F}_\theta$-measurability of Λ_θ for $\theta > t$, implies

$$\mathbb{P}(\tau > \theta | \mathscr{F}_t) = \mathbb{E}\left[\mathbb{P}(\tau > \theta | \mathscr{F}_\theta) | \mathscr{F}_t\right] = \mathbb{E}\left[\exp\left(-\Lambda_\theta\right) | \mathscr{F}_t\right]$$

and the result follows. In particular, we conclude that

$$Z_t = \mathbb{P}(\tau > t | \mathscr{F}_t) = \mathbb{P}(\tau > t | \mathscr{F}_\infty). \tag{2.11}$$

Hence the Azéma supermartingale is decreasing, positive and continuous. □

Corollary 2.26 *The process M defined as $M_t := A_t - \Lambda_{t\wedge\tau}$ is a $\mathbb{G}$-martingale.*

Proof This is an immediate application of Proposition 2.15 and the above lemma since, in a Cox model,

$$\int_0^t (1 - A_{s-})\frac{1}{Z_{s-}} dA_s^p = \int_0^{t\wedge\tau} \exp(\Lambda_s) d(1 - \exp(-\Lambda_s)) = \Lambda_{t\wedge\tau}.$$

□

In a Cox model, Lemma 2.9 and Corollary 2.10 can be written as follow for r.v.'s of a particular form.

Lemma 2.27 *(a) Let X be an integrable $\mathscr{F}_T$-measurable r.v. Then, for $t \le T$*

$$\mathbb{E}\left[X \mathbb{1}_{\{T<\tau\}} | \mathscr{G}_t\right] = \mathbb{1}_{\{\tau>t\}} \mathbb{E}\left[X e^{-(\Lambda_T - \Lambda_t)} \middle| \mathscr{F}_t\right]. \tag{2.12}$$

(b) Let h be a bounded $\mathbb{F}$-predictable process. Then, for $t \le T$

$$\mathbb{E}\left[h_\tau \mathbb{1}_{\{\tau<T\}} | \mathscr{G}_t\right] = h_\tau \mathbb{1}_{\{\tau\le t\}} + \mathbb{1}_{\{\tau>t\}} e^{\Lambda_t} \mathbb{E}\left[\int_t^T h_u \lambda_u e^{-\Lambda_u} du \middle| \mathscr{F}_t\right]. \tag{2.13}$$

The immersion between $\mathbb{F}$ and $\mathbb{G}$ (see Definition 1.20) holds in a Cox model.

Lemma 2.28 *Let τ be a default time in a Cox model. Then $\mathbb{F}$ is immersed in $\mathbb{G}$, i.e., each $\mathbb{F}$-martingale is a $\mathbb{G}$-martingale.*

Proof Since Θ is independent of $\mathscr{F}_\infty$, by Proposition 1.21, we obtain that any $\mathbb{F}$-martingale Y is an $\mathbb{F}^{\sigma(\Theta)} := \mathbb{F} \vee \sigma(\Theta)$-martingale. Since $\mathbb{F} \subset \mathbb{G} \subset \mathbb{F}^{\sigma(\Theta)}$, it follows from Proposition 1.24 that Y is a $\mathbb{G}$-martingale. □

Remark 2.29 (a) We shall see in Sect. 3.2, Lemma 3.8, that the immersion in a Cox model also follows by (2.11).
(b) Immersion has important implications regarding the PRP stability under filtration enlargement. See Sect. 3.2.3 for progressive enlargement of filtration setting and also Proposition 1.23.

2.3.3 Generalisation of Cox Model

Instead of absolutely continuous process Λ we consider an increasing $\mathbb{F}$-adapted process Γ. We emphasize that not only do we not assume that Γ is absolutely continuous, we are even interested in the case where Γ fails to be continuous. Similarly to before, the default time τ is defined as the first time when an increasing process Γ is above the random level Θ, i.e., $\tau := \inf\{t \ge 0 : \Gamma_t \ge \Theta\}$. An analogous argument to previously yields $Z = e^{-\Gamma}$. However, since Γ can fail to be predictable, the $\mathbb{G}$-compensator of A is no longer equal to Γ, as we now demonstrate in an example.

Example 2.30 Let Γ be a compound Poisson process with positive jumps given by $\Gamma_t = \sum_{n=1}^{N_t} Y_n$ where N is a Poisson process with intensity λ and $(Y_n)_{n\geq 1}$ are positive random variables, i.i.d. and independent from N, and let $\mathbb{F}$ be the natural filtration of Γ. Assume that τ is constructed as above.

For $\psi := \int_0^\infty (1 - e^{-y}) F(dy)$ where F is the cumulative distribution function of Y_1, the process $(\mu_t := e^{-\Gamma_t + t\lambda\psi}, t \geq 0)$ is an $\mathbb{F}$-martingale. Then, from $Z_t = \mu_t e^{-t\lambda\psi}$, using integration by parts, one deduces that

$$dZ_t = e^{-t\lambda\psi} d\mu_t - e^{-t\lambda\psi} \mu_t \lambda\psi dt$$

which provides the Doob–Meyer decomposition of Z. It follows, from Proposition 2.15, that

$$M_t := \mathbb{1}_{\{\tau \leq t\}} - (t \wedge \tau)\lambda\psi$$

is an $\mathbb{F}$-martingale, in particular the $\mathbb{F}$-intensity rate of τ is $\lambda\psi$.

2.4 Compensators in a Two Defaults Setting

In this section, in order to underline the role of the filtration in the computations of the compensator, the simplest model with two random times τ_1 and τ_2 is presented. The similar methodology, with more complex computations, can be developed for several default times.

Denote by $\mathbb{A}^i$ the natural filtration of the process $A^i := \mathbb{1}_{[\![\tau_i, \infty[\![}$ for $i = 1, 2$ and by $\mathbb{G}$ the filtration $\mathbb{G} := \mathbb{A}^1 \vee \mathbb{A}^2$. Assume that the pair (τ_1, τ_2) has non-atomic law and its survival probability function $G : \mathbb{R}^+ \times \mathbb{R}^+ \to \mathbb{R}^+$, defined as $G(t, s) = \mathbb{P}(\tau_1 > t, \tau_2 > s)$, is positive and continuously differentiable in both variables. Note that $G(t, 0) = \mathbb{P}(\tau_1 > t)$ is the survival probability of τ_1. We denote by $\partial_i G$ the first partial derivative w.r.t. the ith variable and $\partial_{ij} G$ the second partial derivative w.r.t. the ith and jth variables for $i, j \in \{1, 2\}$.

Proposition 2.31 *(a) The $\mathbb{A}^1$-compensator of τ_1 equals $-\ln G(t \wedge \tau, 0)$, i.e., the process $A_t^1 + \ln G(t \wedge \tau, 0)$ is a $\mathbb{A}^1$-martingale. Equivalently, $\frac{-\partial_1 G(t,0)}{G(t,0)}$ is the intensity rate of τ_1.*
(b) The $\mathbb{G}$-compensator of τ_1 equals $\int_0^t (1 - A_s^1)\widehat{\lambda}_s ds$, where

$$\widehat{\lambda}_t = -\mathbb{1}_{\{t \leq \tau_2\}} \frac{\partial_1 G(t, t)}{G(t, t)} - \mathbb{1}_{\{\tau_2 \leq t\}} \frac{\partial_{12} G(t, \tau_2)}{\partial_2 G(t, \tau_2)}$$

is $\mathbb{A}^2$-intensity rate of τ_1. Equivalently the process $A_t^1 - \int_0^t (1 - A_s)\widehat{\lambda}_s ds$ is a $\mathbb{G}$-martingale.

Proof Assertion (a) is a straightforward consequence of Proposition 2.4.

(b) In order to apply Proposition 2.15 let us first compute the Doob–Meyer decomposition of the supermartingale $Z := {}^{o,\mathbb{A}^2}(\mathbb{1}_{[\![0,\tau_1[\![})$. By Lemma 2.9, one has that

$$Z_t = A_t^2 \mathbb{P}(\tau_1 > t|\sigma(\tau_2)) + (1 - A_t^2)\frac{\mathbb{P}(\tau_1 > t, \tau_2 > t)}{\mathbb{P}(\tau_2 > t)} = A_t^2 h(t, \tau_2) + (1 - A_t^2)\psi(t)$$

where $h(t, v) = \frac{\partial_2 G(t,v)}{\partial_2 G(0,v)}$ and $\psi(t) = \frac{G(t,t)}{G(0,t)}$. Using the integration by parts formula, one obtains

$$dZ_t = (h(t, t) - \psi(t))\, dA_t^2 + \left(A_t^2 \partial_1 h(t, \tau_2) + (1 - A_t^2)\psi'(t)\right) dt.$$

By the assertion (a), the process $M^2 = A^2 + \int_0^{\cdot} (1 - A_t^2)\frac{\partial_2 G(0,t)}{G(0,t)} dt$ is an $\mathbb{A}^2$-martingale and

$$\psi'(t) = (h(t, t) - \psi(t))\frac{\partial_2 G(0, t)}{G(0, t)} + \frac{\partial_1 G(t, t)}{G(0, t)}.$$

It follows that $dZ_t = (h(t, t) - \psi(t))dM_t^2 - dA_t^p$ where

$$dA_t^p = -A_t^2 \partial_1 h(t, \tau_2)dt - (1 - A_t^2)\frac{\partial_1 G(t, t)}{G(0, t)}dt.$$

Finally the form of the $\mathbb{G}$-compensator of τ_1 can be concluded from Proposition 2.4 and the identity $\frac{\partial_1 h(t,\tau_2)}{h(t,\tau_2)} = \frac{\partial_{12} G(t,\tau_2)}{\partial_2 G(t,\tau_2)}$. □

We refer to [194] for related computations.

2.5 Construction of Random Time with Given Intensity

One of the approaches in credit risk is based on the intensity rate, i.e., the knowledge of a process λ such that $A - \int_0^{\cdot} (1 - A_s)\lambda_s ds$ is a martingale. In this formulation, no reference filtration is given, which leads to the following questions. Given a non-negative process λ, is it possible to construct τ such that the previous martingale property holds? The most popular answer is the Cox model (see Sect. 2.3). One can give other constructions as soon as a random time τ, such that the multiplicative decomposition of the Azéma supermartingale is $Ne^{-\Lambda}$, can be constructed. In this section one such a construction is presented; we refer the reader to [133] for a setting where there are infinitely many constructions with the property that the Azéma supermartingale of τ is $Ne^{-\Lambda}$, even in the case $N = 1$. The problem is solved by constructing a random time τ and a probability measure $\mathbb{Q}$ on the product space $\Omega \times \mathbb{R}^+$ such that the Azéma supermartingale of τ under $\mathbb{Q}$ equals $Ne^{-\Lambda}$ and the restriction to $\mathbb{Q}$ to $\mathscr{F}_\infty$ is $\mathbb{P}$. The random variable τ is defined by $\tau((\omega, x)) = x$, and $\mathbb{Q}$ is constructed so that $Y_t(u) := \mathbb{Q}(\tau \leq u|\mathscr{F}_t)$ is a martingale for a fixed u, increasing in u for a fixed t, and $Y_t(t) = \mathbb{Q}(\tau \leq t|\mathscr{F}_t) = 1 - N_t e^{-\Lambda_t} = 1 - Z_t$.

Proposition 2.32 *We assume that N and Λ are continuous and N is positive. Let $0 < u < \infty$ be fixed and consider the process $Y(u)$ defined for $t \in [u, \infty)$ by*

$$Y_t(u) := (1 - Z_t)\exp\left\{-\int_u^t \frac{Z_s}{1 - Z_s} d\Lambda_s\right\}.$$

Then, for any u, the process $Y(u) = (Y_t(u), u \leq t \leq \infty)$ is a uniformly integrable $(\mathbb{F}, \mathbb{P})$-martingale and, for fixed t, the family $(Y_t(u), u \leq t)$ is increasing in u.

Proof Applying the integration by parts formula for $t \geq u$ one gets

$$dY_t(u) = -\exp\left\{-\int_u^t \frac{Z_s}{1 - Z_s} d\Lambda_s\right\} e^{-\Lambda_t} dN_t.$$

Therefore, $Y(u)$ is an $(\mathbb{F}, \mathbb{P})$-local martingale on $[u, \infty)$. Being clearly positive and bounded by 1, it is a uniformly integrable martingale on $[u, \infty)$. □

2.6 Dynamics of Prices

The goal of this section is to give the dynamics of prices of some contingent claims. We assume that the probability measure $\mathbb{P}$ is the pricing measure, i.e., discounted prices are $(\mathbb{G}, \mathbb{P})$-local martingales. We assume that Z is positive, decreasing and continuous and condition **(A)** holds. Denote by Λ the process $\Lambda := -\ln Z$. Then, by Proposition 2.18 (c), the process $\Upsilon := (1 - A)\exp(\Lambda)$ is a $\mathbb{G}$-martingale which satisfies $\Upsilon = 1 - \Upsilon_- \cdot M$. For discounting, the $\mathbb{F}$-adapted interest rate process r and $R_t := \int_0^t r_s ds$ are used.

A **defaultable zero-coupon bond** of maturity T pays one monetary unit at time T, if the default has not occurred before T, and its price at time t is denoted $D_t(T)$.

Proposition 2.33 *Let m^Λ be the $\mathbb{F}$-martingale $m_t^\Lambda := \mathbb{E}_\mathbb{P}\left[\exp(-(R_T + \Lambda_T))|\mathcal{F}_t\right]$. Then the price of the defaultable zero-coupon bond has the following dynamics*

$$dD_t(T) = -D_{t-}(T)dM_t + \Upsilon_{t-}\exp(R_t)dm_t^\Lambda + D_t(T)r_t dt$$

Proof By Lemma 2.9 we obtain that the price $D_t(T)$ of a defaultable zero-coupon bond with maturity T is

$$\begin{aligned} D_t(T) &= \mathbb{E}_\mathbb{P}\left[\mathbb{1}_{\{T<\tau\}}\exp\left(-(R_T - R_t)\right) \middle| \mathcal{G}_t\right] \\ &= \mathbb{1}_{\{\tau>t\}}\mathbb{E}_\mathbb{P}\left[\exp\left(-(R_T + \Lambda_T - R_t - \Lambda_t)\right) \middle| \mathcal{F}_t\right] = m_t^\Lambda \Upsilon_t \exp(R_t). \end{aligned}$$

Note that condition **(A)** implies that the process $[m^K, \Upsilon]$ is constant, since Υ is purely discontinuous martingale with a single jump at τ. Then, by integration by parts,

$$\begin{aligned} dD_t(T) &= -m^\Lambda_{t-}\Upsilon_{t-}\exp(R_t)dM_t + \Upsilon_{t-}\exp(R_t)dm^\Lambda_t + m^\Lambda_t\Upsilon_t\exp(R_t)dR_t \\ &= -D_{t-}(T)dM_t + \Upsilon_{t-}\exp(R_t)dm^\Lambda_t + D_t(T)r_t dt\,. \end{aligned}$$

□

Corollary 2.34 *Moreover if Λ is absolutely continuous, i.e., $\Lambda_t = \int_0^t \lambda_s ds$, and both λ and r are deterministic processes, we have*

$$dD_t(T) = -D_{t-}(T)dM_t + D_t(T)r_t dt\,.$$

Proof In the particular case where λ and r are deterministic, $dm^\Lambda_t = 0$ since m^Λ is a constant process, namely $m^\Lambda = \exp(-(R_T + \Lambda_T))$ and the result follows. □

Let K be a given $\mathbb{F}$-predictable process. A claim with **recovery payment at maturity** is a contract which pays K_τ at date T if $\tau \le T$ and there is no payment in the case $\tau > T$. Its price at time t, for $t \le T$ is denoted V_t. Instead, a claim with **recovery payment at default time** is a contract which pays K_τ at time τ if $\tau \le T$ and there is no payment otherwise. Its price at time t, for $t \le \tau$ is denoted U_t.

Proposition 2.35 *Let m^K be the $\mathbb{F}$-martingale $m^K_t := \mathbb{E}_\mathbb{P}\left[\int_0^T K_u dA^p_u|\mathscr{F}_t\right]$ for a given $\mathbb{F}$-predictable process K. Assume that $r = 0$. Then:*
(a) The process V has dynamics

$$V = V_0 + (K - V_-)\bullet M + \Upsilon_- \bullet m^K.$$

(b) The process U has dynamics

$$U = U_0 - U_- \bullet M + \Upsilon_- \bullet m^K - \Upsilon_- K \bullet A^p.$$

Proof (a) An immediate application of Corollary 2.10 (b) shows that the price at time t of a claim with recovery payment at maturity is

$$\begin{aligned} V_t &:= \mathbb{E}_\mathbb{P}\left[K_\tau 1\!\!1_{\{\tau\le T\}}|\mathscr{G}_t\right] = K_\tau 1\!\!1_{\{\tau\le t\}} + 1\!\!1_{\{t<\tau\}}\frac{1}{Z_t}\mathbb{E}_\mathbb{P}\left[\int_t^T K_u dA^p_u\middle|\mathscr{F}_t\right] \\ &= \int_0^t K_u dA_u + \Upsilon_t\left(-\int_0^t K_u dA^p_u + m^K_t\right). \end{aligned}$$

By integration by parts we obtain

$$\Upsilon m^K = \Upsilon_- \bullet m^K + m^K_t \bullet \Upsilon + [m^K, \Upsilon].$$

As in the proof of Proposition 2.33, the process $[m^K, \Upsilon]$ is constant. Therefore, applying integration by parts again, we deduce that

$$V = V_0 + (K - V_-)\bullet M + \Upsilon_- \bullet m^K.$$

(b) Similarly as for V we derive that

$$U_t = \mathbb{1}_{\{t<\tau\}}\mathbb{E}_{\mathbb{P}}\left[K_\tau \mathbb{1}_{\{t<\tau\le T\}}|\mathcal{G}_t\right] = \mathbb{1}_{\{t<\tau\}}\frac{1}{Z_t}\mathbb{E}_{\mathbb{P}}\left[\int_t^T K_u dA_u^p \middle| \mathcal{F}_t\right].$$

By applying integration by parts, we conclude that the dynamics of U are

$$U = U_0 - U_- \bullet M + \Upsilon_- \bullet m^K - \Upsilon_- K \bullet A^p.$$

□

Remark 2.36 (a) Note that, from the definition, the process V is a $\mathbb{G}$-martingale. Therefore Proposition 2.35 (a) implies that m^K stopped at τ is a $\mathbb{G}$-martingale. We refer the reader to Sect. 5.5 where random times with this property are studied.
(b) Note that $U + \Upsilon_- K \bullet A = V$ is a $\mathbb{G}$-martingale. Assume that the $\mathbb{F}$-intensity rate of τ exists and is denoted by λ. Therefore the process $U_t + \int_0^{t\wedge\tau} K_s\lambda_s ds$ is a $\mathbb{G}$-martingale. The quantity $K_t\lambda_t$ which appears can be interpreted as a dividend K_t paid at rate λ_t, or with probability $\lambda_t dt = \mathbb{P}(t < \tau < t+dt|\mathcal{F}_t)/\mathbb{P}(t < \tau|\mathcal{F}_t)$.

The next proposition concerns pricing and hedging a **defaultable call option** in a Cox model.

Proposition 2.37 *Assume that $\mathbb{F}$ is the natural filtration of a Brownian motion B and τ is a default time in a Cox model such that the process λ is deterministic. Let $r = 0$ and S satisfy $dS_t = S_t\sigma dB_t$, where σ is a constant. Then a defaultable call option, with payoff $\mathbb{1}_{\{T<\tau\}}(S_T - K)^+$, can be perfectly hedged by investing an amount equal to the price of this option in the defaultable bond.*

Proof By Lemma 2.9, the price of a defaultable call option is

$$C_t = \mathbb{E}_{\mathbb{P}}\left[\mathbb{1}_{\{T<\tau\}}(S_T - K)^+|\mathcal{G}_t\right] = \mathbb{1}_{\{t<\tau\}}\frac{1}{Z_t}\mathbb{E}_{\mathbb{P}}\left[Z_T(S_T - K)^+|\mathcal{F}_t\right] = \Upsilon_t m_t^S$$

where $m_t^S := \mathbb{E}_{\mathbb{P}}\left[Z_T(S_T - K)^+|\mathcal{F}_t\right]$. In the particular case where λ is deterministic,

$$m_t^S = e^{-\Lambda_T}\mathbb{E}_{\mathbb{P}}\left[(S_T - K)^+|\mathcal{F}_t\right] = e^{-\Lambda_T}C_t^S$$

where C^S the price of a call in the Black and Scholes model, is of the form $C_t^S = C^S(t, S_t)$, hence

$$C_t = \Upsilon_t e^{-\Lambda_T}C_t^S = D_t(T)C_t^S.$$

Using the continuity of C_t^S and the fact that $dC_t^S = \Delta_t dS_t$ where Δ_t is the Delta-hedge ($\Delta_t = \partial_y C^S(t, S_t)$), we deduce that

$$\begin{aligned} dC_t &= e^{-\Lambda_T}(\Upsilon_t dC_t^S + C_t^S d\Upsilon_t) = e^{-\Lambda_T}(\Upsilon_t\Delta_t dS_t - C_t^S\Upsilon_{t-}dM_t) \\ &= e^{-\Lambda_T}(\Upsilon_t\Delta_t dS_t - C_t^S\Upsilon_{t-}dM_t). \end{aligned}$$

Therefore, since by Corollary 2.34 $dD_t(T) = -e^{-\Lambda_T}\Upsilon_{t-}dM_t$, we obtain

$$dC_t = e^{-\Lambda_T}\Upsilon_t\Delta_t dS_t - C_t^S dD_t(T) = e^{-\Lambda_T}\Upsilon_t\Delta_t dS_t + \frac{C_{t-}}{D_{t-}(T)}dD_t(T),$$

hence a hedging strategy consists of holding $\frac{C_{t-}}{D_{t-}(T)}$ defaultable zero-coupon bonds and $e^{-\Lambda_T}\Upsilon_{t-}\Delta_t$ risky asset S. Note that, on $\{t < \tau\}$, one has $C_{t-} = C_t$. □

The result obtained in Proposition 2.37 can be generalized to the case of stochastic intensity. See the so-called balance condition in [35].

2.7 Bibliographic Notes

The basic case studied in Sect. 2.1 is presented in Brémaud [44] and Dellacherie [71, 72]. Dellacherie and Meyer [76, Chap. IV, paragraph 107] consider also the filtration $\mathbb{A}^* = (\mathscr{A}_t^*, t \geq 0)$ where the σ-field $\mathscr{A}_t^*$ is generated by $\tau \wedge t$ and contains the set $\{\tau \geq t\}$ which cannot be split into two non-null sets from $\mathscr{A}_t^*$ (thus the set $\{\tau \geq t\}$ is an atom of $\mathscr{A}_t^*$). The filtration $\mathbb{A}^*$ is not right-continuous: $\mathscr{A}_{t+}^*$ is obtained by splitting the atom $\{\tau \geq t\}$ into $\{\tau = t\}$ and $\{\tau > t\}$. Setting $\mathscr{A}_t = \mathscr{A}_{t+}^*$, the random time τ is an $\mathbb{A}$-stopping time, but is not an $\mathbb{A}^*$-stopping time. Note that indeed $\mathbb{A}$ is the natural filtration of the process $\mathbb{1}_{[\![\tau,\infty[\![}$. It is proved that any $\mathbb{A}^*$-stopping time is predictable and that, if the law of τ is atomic and not degenerated, then τ is $\mathbb{A}$-accessible and not $\mathbb{A}$-predictable.

The predictable representation property is important in a financial framework, and some results (in particular the case presented in Sect. 2.1.2) can be found in Chou and Meyer [56].

The paper of Herdegen and Herrman [113] contains a systematic study of the case presented in the first section of this chapter, in a very general setting (no specific assumption on the regularity of the law of τ).

A deep study on extension of Proposition 2.11 (b) to optional processes can be found in Song [201].

A chapter in Protter [189] is devoted to the study of compensators. Janson et al. [126] and Zeng [220] present conditions which ensure that the compensator of A is absolutely continuous w.r.t. the Lebesgue measure. Their approach is based on the Ethier–Kurtz criterion. See Janson et al. [126] for an extension of Proposition 2.23 to absolutely continuous compensators.

Compensators can be computed using the *Laplacien approché* methodology (see Dellacherie and Meyer [79, VII, 22] and Dellacherie [74]), also called Aven's lemma in credit risk [23] (see more details in Zeng [220]). As we shall see in Chap. 5, compensators appear in many places for progressive enlargement framework. See also Coculescu [60] for some examples of compensators of default times and Last and Brandt [166] for the more general case of marked point processes.

The intensity based model was introduced in Jarrow and Turnbull [128], and Jarrow et al. [127] (see also Bielecki and Rutkowski [39]). The problem which appear in Sect. 2.5 is studied in Gapeev et al. [103], Jeanblanc and Song [133, 134] and in Li and Rutkowski [168]. It is proven in Song [200] that, for a given supermartingale valued in [0, 1] with multiplicative decomposition $Ne^{-\Lambda}$, there exist various ways to construct τ so that $Z = Ne^{-\Lambda}$, which implies that Λ^τ is the compensator of A. Some constructions lead to immersion, the others do not.

A reduced form approach is presented in Bielecki et al. [36, 37], Elliott et al. [89] and Kusuoka [163] among others. More information on pricing defaultable claims can be found in Bielecki et al. [38]. General presentations of modeling credit risk are done by Giesecke [105] and Bélanger et al. [32]. In particular, Cox models are used in a great number of studies (see, e.g., Lando [164]).

Structural models are models for default time, in which the default time is defined as a stopping time in the given filtration. These models do not involve enlargement of filtration and are not presented here. Guo and Zheng [111] present some explicit computation of compensators for hitting times and Okhrati et al. [187] study compensators of processes of the form $g(t, X_t)\mathbb{1}_{\{\tau>t\}}$ where τ is a hitting time for a Lévy process X. The structural approach faced some difficulties. One of them is that, if the filtration taken into account satisfies condition (**C**), the random time τ is predictable, hence prices of defaultable zero coupon must go to 0 before τ, and this fact is not observed in the data. Another difficulty is to compute the conditional law of τ, especially in the case of multidimensional default times (see Blanchet and Patras [42]).

2.8 Exercises

Exercise 2.1 Let B be a Brownian motion and $\tau = \inf\{t : B_t = a\}$. Find the $\mathbb{F}^B$-compensator of τ and the $\mathbb{A}$-compensator of τ, where $\mathbb{A}$ is the natural filtration of the process A.

Exercise 2.2 We are in the setting of Sect. 2.1 and we assume that F is differentiable, $F < 1$ and $F(\infty) = 1$. Set $\Lambda(t) = \int_0^t \frac{dF(s)}{1-F(s)}$.
(a) Let $h : \mathbb{R}_+ \to \mathbb{R}$ be a (bounded) Borel measurable function. Prove that the process

$$Y_t := \exp\left(\mathbb{1}_{\{\tau\le t\}}h(\tau)\right) - \int_0^{t\wedge\tau} (e^{h(u)} - 1)\, d\Lambda(u)$$

is an $\mathbb{A}$-martingale. Find an $\mathbb{A}$-predictable process φ such that $dY_t = \varphi_t dM_t$.
(b) Let $h : \mathbb{R}_+ \to \mathbb{R}$ be a non-negative Borel measurable function such that the random variable $h(\tau)$ is integrable. Prove that the process

$$Y_t := (1 + \mathbb{1}_{\{\tau\le t\}}h(\tau)) \exp\left(-\int_0^{t\wedge\tau} h(u)\, d\Lambda(u)\right).$$

is an $\mathbb{A}$-martingale. Find an $\mathbb{A}$-predictable process φ such that $dY_t = \varphi_t dM_t$. Give a condition on h so that Y is positive. In that case, find an $\mathbb{A}$-predictable process ψ such that $dY_t = Y_{t-}\psi_t dM_t$.

Exercise 2.3 Assume that

$$dS_t = S_t(rdt + \sigma dB_t), \quad S_0 = 1$$

where B is a Brownian motion and let $\tau = \inf\{t : S_t \leq \alpha\}$, with $\alpha < 1$. Define $\mathbb{H} = (\mathscr{H}_t, t \geq 0)$ as the filtration generated by the observations of S at given times $t_1, \dots, t_n$, i.e., for $t \in [t_n, t_{n+1})$, that is, $\mathscr{H}_t = \sigma(S_s, s \leq t_n)$ for $t_n \leq t < t_{n+1}$. Compute the $\mathbb{H}$-intensity rate of τ.

Exercise 2.4 (a) Prove that, in a Cox model, τ is independent of $\mathscr{F}_\infty$ if and only if λ is a deterministic function.
(b) Prove that, in general, in a Cox model, $\mathbb{A}$ is not immersed in $\mathbb{G}$. Prove that, if λ is deterministic, $\mathbb{A}$ is immersed in $\mathbb{G}$.

Chapter 3
Immersion

Given a pair of filtrations $\mathbb{F}$ and $\mathbb{G}$ such that $\mathbb{F} \subset \mathbb{G}$, we study the particular case when all $\mathbb{F}$-martingales are $\mathbb{G}$-martingales. This property is known as immersion or hypothesis $\mathscr{H}$ between $\mathbb{F}$ and $\mathbb{G}$.

We give a complete characterization of filtrations $\mathbb{F}$ and $\mathbb{G}$ such that immersion holds. We assume neither that $\mathbb{G}$ is a progressive enlargement nor an initial enlargement of $\mathbb{F}$. Immersion strongly relies on the underlying probability measure, and is not stable under an equivalent change of probability measure. In this direction, we provide some sufficient conditions on the Radon–Nikodym derivative under which immersion holds for an equivalent change of probability measure. Furthermore we present standard examples, paying particular attention to the case where $\mathbb{G}$ is the progressive enlargement of $\mathbb{F}$ with a random time τ.

3.1 Immersion of Filtrations

3.1.1 Definition and Examples

We recall Definition 1.20 (a) given in Chap. 1.

Definition 3.1 Let $\mathbb{F} \subset \mathbb{G}$. The filtration $\mathbb{F}$ is said to be **immersed** in the filtration $\mathbb{G}$ if any $\mathbb{F}$-martingale is a $\mathbb{G}$-martingale. We denote this property by $\mathbb{F} \hookrightarrow \mathbb{G}$.

By localization arguments, it can be proven that immersion holds if and only if every $\mathbb{F}$-local martingale is a $\mathbb{G}$-local martingale or if and only if every $H^1(\mathbb{F})$-martingale is an $H^1(\mathbb{G})$-martingale. Since bounded martingales are dense in $H^1(\mathbb{F})$, one has $\mathbb{F} \hookrightarrow \mathbb{G}$ if and only if every bounded $\mathbb{F}$-martingale is a bounded $\mathbb{G}$-martingale.

Theorem 3.2 *The following assertions are equivalent.*

A. Aksamit and M. Jeanblanc, *Enlargement of Filtration with Finance in View*, SpringerBriefs in Quantitative Finance,
https://doi.org/10.1007/978-3-319-41255-9_3

(a) The filtration $\mathbb{F}$ is immersed in the filtration $\mathbb{G}$.
(b) For each $t \geq 0$, the σ-fields $\mathscr{F}_\infty$ and $\mathscr{G}_t$ are conditionally independent given $\mathscr{F}_t$, i.e., for each $t \geq 0$, one has $\mathbb{E}[G_t F|\mathscr{F}_t] = \mathbb{E}[G_t|\mathscr{F}_t]\mathbb{E}[F|\mathscr{F}_t]$ for every $G_t \in L^2(\mathscr{G}_t)$ and $F \in L^2(\mathscr{F}_\infty)$.
(c) For every $t \geq 0$ and every $G_t \in L^1(\mathscr{G}_t)$ one has $\mathbb{E}[G_t|\mathscr{F}_\infty] = \mathbb{E}[G_t|\mathscr{F}_t]$.
(d) For every $t \geq 0$ and every $F \in L^1(\mathscr{F}_\infty)$ one has $\mathbb{E}[F|\mathscr{G}_t] = \mathbb{E}[F|\mathscr{F}_t]$.
Furthermore, if $\mathbb{F} \hookrightarrow \mathbb{G}$, then, for every $t \geq 0$, one has $\mathscr{G}_t \cap \mathscr{F}_\infty = \mathscr{F}_t$.

Proof First we prove (a) $\Rightarrow$ (b). Let $F \in L^2(\mathscr{F}_\infty)$ and assume that (a) is satisfied. This implies that the $\mathbb{F}$-martingale $\mathbb{E}[F|\mathscr{F}_t]$ is a $\mathbb{G}$-martingale with terminal value F, hence $\mathbb{E}[F|\mathscr{F}_t] = \mathbb{E}[F|\mathscr{G}_t]$. It follows that for any $t \geq 0$ and any $G_t \in L^2(\mathscr{G}_t)$

$$\mathbb{E}[FG_t|\mathscr{F}_t] = \mathbb{E}[G_t\mathbb{E}[F|\mathscr{G}_t]|\mathscr{F}_t] \stackrel{(a)}{=} \mathbb{E}[G_t\mathbb{E}[F|\mathscr{F}_t]|\mathscr{F}_t] = \mathbb{E}[G_t|\mathscr{F}_t]\mathbb{E}[F|\mathscr{F}_t]$$

which is exactly (b).

To prove (b) $\Rightarrow$ (c), let $F \in L^2(\mathscr{F}_\infty)$ and $G_t \in L^2(\mathscr{G}_t)$. Under (b),

$$\begin{aligned}\mathbb{E}[F\mathbb{E}[G_t|\mathscr{F}_t]] &= \mathbb{E}[\mathbb{E}[F|\mathscr{F}_t]\mathbb{E}[G_t|\mathscr{F}_t]] \stackrel{(b)}{=} \mathbb{E}[\mathbb{E}[FG_t|\mathscr{F}_t]] \\ &= \mathbb{E}[FG_t] = \mathbb{E}[F\mathbb{E}[G_t|\mathscr{F}_\infty]],\end{aligned}$$

thus $\mathbb{E}[G_t|\mathscr{F}_\infty] = \mathbb{E}[G_t|\mathscr{F}_t]$. By Lebesgue's monotone convergence theorem the equality is extended to $G_t \in L^1(\mathscr{G}_t)$.

Now, we prove (c) $\Rightarrow$ (d). Let $F \in L^2(\mathscr{F}_\infty)$ and $G_t \in L^2(\mathscr{G}_t)$. If (c) holds, then,

$$\mathbb{E}[G_t\mathbb{E}[F|\mathscr{F}_t]] = \mathbb{E}[F\mathbb{E}[G_t|\mathscr{F}_t]] \stackrel{(c)}{=} \mathbb{E}[F\mathbb{E}[G_t|\mathscr{F}_\infty]] = \mathbb{E}[FG_t] = \mathbb{E}[G_t\mathbb{E}[F|\mathscr{G}_t]],$$

which implies $\mathbb{E}[F|\mathscr{G}_t] = \mathbb{E}[F|\mathscr{F}_t]$. By Lebesgue's monotone convergence theorem the equality is extended to $F \in L^1(\mathscr{F}_\infty)$. Finally, obviously (d) implies (a).

To prove the equality $\mathscr{G}_t \cap \mathscr{F}_\infty = \mathscr{F}_t$, we only have to check that $\mathscr{G}_t \cap \mathscr{F}_\infty \subset \mathscr{F}_t$. For $A \in \mathscr{G}_t \cap \mathscr{F}_\infty$, we have $\mathbb{1}_A = \mathbb{E}[\mathbb{1}_A|\mathscr{F}_\infty] \stackrel{(c)}{=} \mathbb{E}[\mathbb{1}_A|\mathscr{F}_t]$, hence $A \in \mathscr{F}_t$. □

Remark 3.3 (a) If B is an $\mathbb{F}$-Brownian motion and $\mathbb{F} \hookrightarrow \mathbb{G}$, then B is as well a $\mathbb{G}$-Brownian motion. Indeed, for any σ, the process $\exp(\sigma B_t - \frac{1}{2}\sigma^2 t)$ is an $\mathbb{F}$-martingale, hence a $\mathbb{G}$-martingale, and B is a $\mathbb{G}$-Brownian motion. One can also argue this by noticing that the predictable quadratic variation process of a continuous $\mathbb{F}$-martingale is the same in both filtrations $\mathbb{F}$ and $\mathbb{G}$, thus, by Lévy's theorem (see [136, Theorem 1.4.1.2]), B is a $\mathbb{G}$-Brownian motion.
(b) If for any t, $\mathscr{F}_t \subset \mathscr{G}_t \subset \mathscr{F}_\infty$, and if $\mathbb{F} \hookrightarrow \mathbb{G}$, then the last assertion of Theorem 3.2 implies that $\mathbb{G}$ equals $\mathbb{F}$.

In Sects. 1.2.4 and 2.3 examples of immersion were already given. We present now other ones.

Example 3.4 Let, for $n > 1$, $B = (B^{(1)}, \dots, B^{(n)})$ be an n-dimensional Brownian motion. Let R be the Euclidean norm of B, i.e., $R_t = ||B_t||$, then R is an n-dimensional Bessel process. By Itô's formula, the decomposition of the semimartingale R is

$$R_t = \beta_t + \frac{n-1}{2}\int_0^t \frac{ds}{R_s},$$

where β is the $\mathbb{F}^B$-Brownian motion given by

$$\beta_t = \sum_{i=1}^{n} \int_0^t \frac{B_s^{(i)} dB_s^{(i)}}{R_s}. \tag{3.1}$$

Clearly β is $\mathbb{F}^R$-adapted. On the other hand, R is $\mathbb{F}^\beta$-adapted, by uniqueness of the strong solution of the associated SDE. Thus we conclude that $\mathbb{F}^\beta = \mathbb{F}^R$ and β has the PRP in $\mathbb{F}^R$. It then follows that $\mathbb{F}^R$ is immersed in $\mathbb{F}^B$ since any $\mathbb{F}^R$-martingale can be written as a stochastic integral w.r.t. β and β is a Brownian motion in both filtrations $\mathbb{F}^R$ and $\mathbb{F}^B$.

Example 3.5 Let B be a Brownian motion and β the Brownian motion defined as $\beta_t = \int_0^t \mathrm{sgn}(B_s)dB_s$ (see Exercise 1.3). Then, from Tanaka's formula (see, e.g., [136, Sect. 4.1.8]), $|B| = \beta + L$ where L is the local time of B at level 0, and one can deduce that $\mathbb{F}^\beta$ is equal to $\mathbb{F}^{|B|}$. Any $\mathbb{F}^\beta$-martingale being a stochastic integral w.r.t. β (hence w.r.t. B) is an $\mathbb{F}^B$-martingale, and $\mathbb{F}^\beta$ is immersed in $\mathbb{F}^B$.

3.1.2 Change of Probability Measure

Immersion strongly depends on the probability measure, and in particular, is not stable under a change of probability measure. Assuming that $\mathbb{F} \hookrightarrow \mathbb{G}$ under $\mathbb{P}$, and that $\mathbb{Q}$ is equivalent to $\mathbb{P}$ on $\mathbb{G}$, any $(\mathbb{F}, \mathbb{Q})$-martingale is a $(\mathbb{G}, \mathbb{Q})$-semimartingale, and we now give its decomposition. We also provide conditions on the change of probability measure which imply that immersion is preserved.

Proposition 3.6 *Assume that $\mathbb{F} \hookrightarrow \mathbb{G}$ under $\mathbb{P}$. Let $\mathbb{Q}$ be equivalent to $\mathbb{P}$ on $\mathcal{G}_\infty$ and for $t \in [0, \infty)$, denote $\frac{d\mathbb{Q}}{d\mathbb{P}}|_{\mathcal{G}_t} = L_t$ and $\frac{d\mathbb{Q}}{d\mathbb{P}}|_{\mathcal{F}_t} = \ell_t = \mathbb{E}[L_t|\mathcal{F}_t]$.*
(a) If X is an $(\mathbb{F}, \mathbb{Q})$-local martingale, the process

$$X_t + \int_0^t \frac{L_{s-}}{L_s}\left(\frac{1}{\ell_{s-}}d[X, \ell]_s - \frac{1}{L_{s-}}d[X, L]_s\right) = X_t + \int_0^t \frac{1}{Y_{s-}}d[X, Y]_s$$

where $Y = \ell/L$, is a $(\mathbb{G}, \mathbb{Q})$-local martingale.
(b) $\mathbb{F} \hookrightarrow \mathbb{G}$ under $\mathbb{Q}$ if and only if, for any $\zeta \in L^1(\mathcal{F}_\infty)$

$$\frac{\mathbb{E}_\mathbb{P}[\zeta L_\infty|\mathcal{G}_t]}{L_t} = \frac{\mathbb{E}_\mathbb{P}[\zeta \ell_\infty|\mathcal{F}_t]}{\ell_t}.$$

(c) If L is assumed to be $\mathbb{F}$-adapted, then $\mathbb{F} \hookrightarrow \mathbb{G}$ under $\mathbb{Q}$. Furthermore, in the case of progressive enlargement $\mathbb{G}$ of the filtration $\mathbb{F}$ with a random time τ, the Azéma supermartingales of τ under $\mathbb{P}$ and $\mathbb{Q}$ are equal.

Proof (a) Given an $(\mathbb{F}, \mathbb{Q})$-local martingale X we know, thanks to Girsanov's theorem, that

$$\widetilde{X}_t := X_t - \int_0^t \ell_s d\left[X, \frac{1}{\ell}\right]_s$$

is an $(\mathbb{F}, \mathbb{P})$-local martingale, that remains a $(\mathbb{G}, \mathbb{P})$-local martingale, since $\mathbb{F} \hookrightarrow \mathbb{G}$ under $\mathbb{P}$. We then apply once more Girsanov's theorem to pass from $\mathbb{P}$ to $\mathbb{Q}$ under $\mathbb{G}$ by means of L, so that we obtain that

$$\overline{X}_t: = \widetilde{X}_t - \int_0^t \frac{1}{L_s} d\,[\widetilde{X}, L]_s$$

is a $(\mathbb{G}, \mathbb{Q})$-local martingale. Obviously,

$$\begin{aligned}\overline{X} &= X - \ell \cdot \left[X, \frac{1}{\ell}\right] - \frac{1}{L} \cdot [X, L] + \frac{\ell}{L} \cdot \left[L, \left[X, \frac{1}{\ell}\right]\right] \\ &= X + \ell\left(\frac{\Delta L}{L} - 1\right) \cdot \left[X, \frac{1}{\ell}\right] - \frac{1}{L} \cdot d[X, L] = X - \frac{\ell L_-}{L} \cdot \left[X, \frac{1}{\ell}\right] - \frac{1}{L} \cdot d[X, L].\end{aligned}$$

The result follows by noticing that $\ell \cdot [X, \frac{1}{\ell}] = -\frac{1}{\ell_-} \cdot [X, \ell]$. A simple computation of $[X, Y]$ based on Itô's lemma leads to the last equality.
(b) This is an application of Bayes' formula and condition (d) in Theorem 3.2.
(c) From (b), since $L = \ell$, one has $\mathbb{F} \hookrightarrow \mathbb{G}$ under $\mathbb{Q}$. Furthermore, in the progressively enlarged filtration case, using Bayes' formula, we have:

$$\mathbb{Q}(\tau > t \,|\, \mathscr{F}_t) = \frac{\mathbb{E}_{\mathbb{P}}[L_t 1\!\!1_{\{\tau > t\}} \,|\, \mathscr{F}_t]}{\mathbb{E}_{\mathbb{P}}[L_t \,|\, \mathscr{F}_t]} = \mathbb{P}(\tau > t \,|\, \mathscr{F}_t).$$

□

Example 3.7 Kusuoka [163] presents the following counterexample to the stability of immersion under a change of probability. Let τ_1 and τ_2 be two independent random times which admit densities g_1 and g_2 w.r.t. Lebesgue's measure (i.e., $\mathbb{P}(\tau_i \in ds) = g_i(s)ds$ for $i = 1, 2$) and A^i the associated default processes for $i = 1, 2$. From (2.2), the process $M_t^1 = A_t^1 - \int_0^{t\wedge\tau_1} \lambda_1(s)\, ds$ is an $(\mathbb{A}^1, \mathbb{P})$-martingale, where $\lambda_1(s) = g_1(s)/(\int_s^\infty g_1(u)du)$. Due to the independence of τ_1 and τ_2 under $\mathbb{P}$, immersion under $\mathbb{P}$ between $\mathbb{A}^1$ and $\mathbb{G} := \mathbb{A}^1 \vee \mathbb{A}^2$ holds and M^1 is a $(\mathbb{G}, \mathbb{P})$-martingale. Let us define

$$L_t = 1 + \int_0^t L_{u-} \kappa_u \, dM_u^1$$

for some $\mathbb{G}$-predictable process κ, which is not deterministic, satisfying $\kappa > 1$ and such that L is a u.i. $\mathbb{G}$-martingale. Let $d\mathbb{Q}|_{\mathcal{G}_\infty} = L_\infty \, d\mathbb{P}|_{\mathcal{G}_\infty}$,

$$\widehat{M}_t^1 = A_t^1 - \int_0^{t\wedge\tau_1} \widehat{\lambda}_1(s)\, ds \quad \text{and} \quad \widetilde{M}_t^1 = A_t^1 - \int_0^{t\wedge\tau_1} \lambda_1(s)(1+\kappa_s)\, ds\,,$$

where $\widehat{\lambda}_1(s)ds = \mathbb{Q}(\tau_1 \in ds)/\mathbb{Q}(\tau_1 > s)$ is deterministic. From (2.2), the process $\widehat{M}^1$ is an $(\mathbb{A}^1, \mathbb{Q})$-martingale and $\widetilde{M}^1$ is a $(\mathbb{G}, \mathbb{Q})$-martingale by application of Girsanov's theorem. The uniqueness of the compensator implies that the process $\widehat{M}^1$ is not a $(\mathbb{G}, \mathbb{Q})$-martingale, hence, $\mathbb{A}^1$ is not immersed in $\mathbb{G}$ under $\mathbb{Q}$.

3.2 Immersion for a Progressively Enlarged Filtration

Consider the case where a filtration $\mathbb{F}$ is given as well as a random time τ. As before, we shall work with the process $A := \mathbb{1}_{[\![\tau,\infty[\![}$, the $\mathbb{F}$-Azéma supermartingale $Z := {}^{o,\mathbb{F}}(1-A)$ and the filtration $\mathbb{G} = \mathbb{F} \vee \mathbb{A}$.

3.2.1 Characterization

Lemma 3.8 *The filtration $\mathbb{F}$ is immersed in $\mathbb{G}$ if and only if one of the following equivalent conditions holds:*

$$\forall t,\;\; \mathbb{P}(\tau \le t|\mathcal{F}_\infty) = \mathbb{P}(\tau \le t|\mathcal{F}_t)\,, \tag{3.2}$$

$$\forall t, \forall s \ge t,\;\; \mathbb{P}(\tau \le t|\mathcal{F}_s) = \mathbb{P}(\tau \le t|\mathcal{F}_t)\,, \tag{3.3}$$

$$\forall t,\; \mathbb{P}(\tau < t|\mathcal{F}_\infty) = \mathbb{P}(\tau < t|\mathcal{F}_t)\,. \tag{3.4}$$

Proof First note that, by applying the monotone class theorem, $\mathbb{F} \hookrightarrow \mathbb{G}$ is equivalent to

$$\forall t, \forall s \ge t,\;\; \mathbb{P}(\tau \le t|\mathcal{F}_s) = \mathbb{P}(\tau \le t|\mathcal{F}_\infty) \tag{3.5}$$

which is precisely the condition (b) of Theorem 3.2 in the special case of progressive enlargement of filtration. Then clearly (3.5) implies (3.2). Condition (3.2) implies (3.3) by the tower property and the fact that $\mathcal{F}_s \subset \mathcal{F}_\infty$. Condition (3.3) implies (3.5) since, for all $t \ge 0$, $u \ge t$ and $s \ge t$, by Theorem 1.11, we have

$$\mathbb{P}(\tau \le t|\mathcal{F}_\infty) = \lim_{u\nearrow\infty} \mathbb{P}(\tau \le t|\mathcal{F}_u) = \lim_{u\nearrow\infty} \mathbb{P}(\tau \le t|\mathcal{F}_s) = \mathbb{P}(\tau \le t|\mathcal{F}_s).$$

Equivalence between (3.2) and (3.4) follows on the one hand by

$$\mathbb{P}(\tau < t|\mathcal{F}_\infty) = \lim_{\varepsilon\searrow 0} \mathbb{P}(\tau \le t-\varepsilon|\mathcal{F}_\infty) = \lim_{\varepsilon\searrow 0} \mathbb{P}(\tau \le t-\varepsilon|\mathcal{F}_t) = \mathbb{P}(\tau < t|\mathcal{F}_t)$$

and on the other hand since $\mathbb{F}$ is a right-continuous filtration and

$$\mathbb{P}(\tau \leq t|\mathscr{F}_\infty) = \lim_{\varepsilon \searrow 0} \mathbb{P}(\tau < t+\varepsilon|\mathscr{F}_\infty) = \lim_{\varepsilon \searrow 0} \mathbb{P}(\tau < t+\varepsilon|\mathscr{F}_{t+\varepsilon}) = \mathbb{P}(\tau \leq t|\mathscr{F}_{t+}).$$

□

If $\mathbb{F} \hookrightarrow \mathbb{G}$ holds, then (3.4) implies that the process $\mathbb{P}(\tau \geq t|\mathscr{F}_t)$ is decreasing. In the next proposition a stronger implication of immersion for a progressively enlarged filtration is stated.

Proposition 3.9 *(a) If $\mathbb{F} \hookrightarrow \mathbb{G}$, then $Z = 1 - A^o$, i.e., $m = 1$ where m is introduced in Proposition 1.46 (b).*
(b) If condition **(C)** *or* **(A)** *holds and $\mathbb{F} \hookrightarrow \mathbb{G}$, then the Doob–Meyer decomposition of Z is $Z = 1 - A^p$.*
(c) If $\mathbb{F} \hookrightarrow \mathbb{G}$, Z is positive and condition **(A)** *holds then Z is continuous and the $\mathbb{G}$-martingale M, introduced in (2.10), is $M_t = A_t - \Lambda_{t\wedge\tau}$, where $\Lambda := \frac{1}{Z} \cdot A^p = -\ln Z$.*

Proof (a) Under immersion the $\mathbb{F}$-martingale m is a $\mathbb{G}$-martingale, which is locally bounded (as a BMO-martingale it has bounded jumps). Hence by optional sampling theorem one deduces that $\mathbb{E}[m_{\tau\wedge T_n}] = \mathbb{E}[m_0] = 1$ where (T_n) is a sequence of $\mathbb{F}$-stopping times such that $T_n \nearrow \infty$ and $\mathbb{F}$-martingale $m_{t\wedge T_n}$ is bounded for each n. Then, by Proposition 1.49 (b), we conclude that m is a constant martingale since $\mathbb{E}[[m]_{T_n}] = \mathbb{E}[m_{\tau\wedge T_n}] = 1$ for each n. The result follows by Proposition 1.46 (b).
(b) If condition **(C)** or **(A)** Lemma 1.48 implies that $A^o = A^p$ and the result follows by assertion (a).
(c) Under condition **(A)**, A^p is continuous, thus, applying (b), we obtain that Z is continuous and decreasing. Then, by Proposition 2.15, it is enough to show that $\Lambda = -\ln Z$ which holds from continuity of Z and $\Lambda := \frac{1}{Z} \cdot A^p = -\frac{1}{Z} \cdot Z = -\ln Z$. □

Remark 3.10 (a) A random time such that $Z = 1 - A^o$, or equivalently such that $m = 1$, is called a pseudo-stopping time. The class of pseudo-stopping times is studied in detail in Sect. 5.5.
(b) Note that if $\mathbb{F} \hookrightarrow \mathbb{G}$ and condition **(A)** holds then the process Z is decreasing and continuous. Therefore, the assertion in Proposition 3.9 (c) is in fact a consequence of Proposition 2.18 (c).

3.2.2 $\mathbb{G}$-Martingales Versus $\mathbb{F}$-Martingales

Proposition 3.11 *Let Y be a $\mathbb{G}$-optional process, such that Y_t is integrable for any $t \geq 0$, and*

$$Y = y \mathbb{1}_{[\![0,\tau[\![} + \widehat{y}(\tau) \mathbb{1}_{[\![\tau,\infty[\![}, \tag{3.6}$$

where y is an $\mathbb{F}$-optional process and $(\omega, t, u) \mapsto \widehat{y}_t(\omega, u)$ is an $\mathscr{O}(\mathbb{F}) \otimes \mathscr{B}(\overline{\mathbb{R}}^+)$-measurable function.

Assume that $\mathbb{F}$ is immersed in $\mathbb{G}$ and that
(a) the $\mathbb{F}$-optional projection of Y is an $\mathbb{F}$-martingale,
(b) for any fixed $u \in \mathbb{R}^+$, the process $(\widehat{y}_t(u),\ t \in [u, \infty))$ is an $\mathbb{F}$-martingale.
Then, Y is a $\mathbb{G}$-martingale.

Proof Let $s \le t$. Then,

$$\mathbb{E}[Y_t|\mathscr{G}_s] = \mathbb{E}\left[y_t \mathbb{1}_{\{\tau>t\}}|\mathscr{G}_s\right] + \mathbb{E}\left[\widehat{y}_t(\tau)\mathbb{1}_{\{s<\tau\le t\}}|\mathscr{G}_s\right] + \mathbb{E}\left[\widehat{y}_t(\tau)\mathbb{1}_{\{\tau\le s\}}|\mathscr{G}_s\right]. \quad (3.7)$$

Firstly we shall prove that

$$\mathbb{E}[\widehat{y}_t(\tau)\mathbb{1}_{\{\tau\le s\}}|\mathscr{G}_s] = \mathbb{1}_{\{\tau\le s\}}\widehat{y}_s(\tau). \quad (3.8)$$

Using the usual argument based on the monotone class theorem, it suffices to establish the previous equality for a generator of the $\mathscr{O}(\mathbb{F}) \otimes \mathscr{B}(\overline{\mathbb{R}}^+)$-measurable functions satisfying (b), i.e., for $\widehat{y}_t(u) = h(u)X_t$, where $(X_t, t \ge u)$ is an $\mathbb{F}$-martingale and h is a Borel function. For such $\widehat{y}(u)$, one indeed has

$$\begin{aligned}\mathbb{E}[\widehat{y}_t(\tau)\mathbb{1}_{\{\tau\le s\}}|\mathscr{G}_s] &= \mathbb{E}[X_t h(\tau)\mathbb{1}_{\{\tau\le s\}}|\mathscr{G}_s] = \mathbb{1}_{\{\tau\le s\}}h(\tau)\mathbb{E}[X_t|\mathscr{G}_s]\\ &= \mathbb{1}_{\{\tau\le s\}}h(\tau)\mathbb{E}[X_t|\mathscr{F}_s] = \mathbb{1}_{\{\tau\le s\}}h(\tau)X_s = \mathbb{1}_{\{\tau\le s\}}\widehat{y}_s(\tau)\end{aligned}$$

where the third equality follows from $\mathbb{F} \hookrightarrow \mathbb{G}$.

Secondly, applying Lemma 2.9 to the first two terms in (3.7), we obtain

$$\mathbb{E}[Y_t \mathbb{1}_{\{s<\tau\}}|\mathscr{G}_s] = \mathbb{1}_{\{s<\tau\}}\frac{1}{Z_s}\Big(\mathbb{E}\left[y_t Z_t|\mathscr{F}_s\right] + \mathbb{E}\left[\widehat{y}_t(\tau)\mathbb{1}_{\{s<\tau\le t\}}|\mathscr{F}_s\right]\Big). \quad (3.9)$$

Since ${}^{o,\mathbb{F}}Y$ is an $\mathbb{F}$-martingale, one has

$$\mathbb{E}\left[y_t Z_t + \widehat{y}_t(\tau)\mathbb{1}_{\{\tau\le t\}}|\mathscr{F}_s\right] = \mathbb{E}[Y_t|\mathscr{F}_s] = \mathbb{E}[Y_s|\mathscr{F}_s] = y_s Z_s + \mathbb{E}\left[\widehat{y}_s(\tau)\mathbb{1}_{\{\tau\le s\}}|\mathscr{F}_s\right]$$

from which we derive $\mathbb{E}[y_t Z_t|\mathscr{F}_s]$ and insert it in (3.9) to obtain

$$\mathbb{E}[Y_t \mathbb{1}_{\{s<\tau\}}|\mathscr{G}_s] = \mathbb{1}_{\{s<\tau\}}\frac{1}{Z_s}\Big(y_s Z_s + \mathbb{E}\left[(\widehat{y}_s(\tau) - \widehat{y}_t(\tau))\mathbb{1}_{\{\tau\le s\}}|\mathscr{F}_s\right]\Big).$$

It remains to check that $\mathbb{E}\left[(\widehat{y}_s(\tau) - \widehat{y}_t(\tau))\mathbb{1}_{\{\tau\le s\}}|\mathscr{F}_s\right] = 0$, which follows from

$$\mathbb{E}\left[\widehat{y}_t(\tau)\mathbb{1}_{\{\tau\le s\}}|\mathscr{F}_s\right] = \mathbb{E}\left[\mathbb{E}\left[\widehat{y}_t(\tau)\mathbb{1}_{\{\tau\le s\}}|\mathscr{G}_s\right]|\mathscr{F}_s\right] = \mathbb{E}\left[\widehat{y}_s(\tau)\mathbb{1}_{\{\tau\le s\}}|\mathscr{F}_s\right]$$

where we have used again (3.8). □

3.2.3 Predictable Representation Property

In this subsection we study the propagation of the PRP to a progressively enlarged filtration under immersion. We present a generic result in Theorem 3.12 assuming in particular that condition (**A**) is satisfied. It can be applied to two classical examples of the PRP: $\mathbb{F}$ is a Brownian filtration or $\mathbb{F}$ is a Poisson filtration, and, by mean of a countable family of martingales, in a Lévy filtration (see Di Tella and Engelbert [81, 82]). A general result, established in [135], is recalled in Theorem 3.13.

Theorem 3.12 *Suppose that $(X^1, \ldots, X^d)$ has the PRP in $\mathbb{F}$, i.e., for any $\mathbb{F}$-local martingale Y there exist $\mathbb{F}$-predictable processes $\psi^1, \ldots, \psi^d$ s.t. $Y = Y_0 + \sum_{i=1}^d \psi^i \cdot X^i$. Assume that $\mathbb{F} \hookrightarrow \mathbb{G}$ and that Z is positive and condition* (**A**) *holds. Then, any $\mathbb{G}$-local martingale X admits a representation as*

$$X = X_0 + \sum_{i=1}^{d} \varphi^i \cdot X^i + \gamma \cdot M, \tag{3.10}$$

where φ^i and γ are $\mathbb{G}$-predictable and M is given in (2.10).

Proof We consider $\mathbb{G}$-martingales of the form $X_t := \mathbb{E}[\xi h(\tau)|\mathscr{G}_t]$ for an integrable $\mathscr{F}_\infty$-measurable random variable ξ and a bounded Borel measurable function h. Then, from Lemma 2.9, we have

$$X_t = A_t h(\tau)\mathbb{E}[\xi|\mathscr{G}_t] + (1 - A_t)\frac{1}{Z_t}\mathbb{E}\left[\xi h(\tau)\mathbb{1}_{\{\tau>t\}}\middle|\mathscr{F}_t\right].$$

From immersion $x_t := \mathbb{E}[\xi|\mathscr{F}_t] = \mathbb{E}[\xi|\mathscr{G}_t]$. Defining the $\mathbb{F}$-martingale $z_t := \mathbb{E}\left[\xi h(\tau)\middle|\mathscr{F}_t\right]$, we have that

$$\mathbb{E}\left[\xi h(\tau)\mathbb{1}_{\{\tau>t\}}\middle|\mathscr{F}_t\right] = z_t - x_t \int_{[0,t]} h(u)dA^p_u$$

since, by Theorem 3.2 (b) and the definition of dual predictable projection,

$$\mathbb{E}\left[\xi h(\tau)\mathbb{1}_{\{\tau\le t\}}\middle|\mathscr{F}_t\right] = \mathbb{E}\left[\xi|\mathscr{F}_t\right]\mathbb{E}\left[h(\tau)\mathbb{1}_{\{\tau\le t\}}\middle|\mathscr{F}_t\right] = x_t \int_{[0,t]} h(u)dA^p_u.$$

Under the assumptions on Z, Proposition 3.9 (c) and Proposition 2.18 (c) imply that the process $\Upsilon := (1-A)\frac{1}{Z}$ is a $\mathbb{G}$-martingale satisfying $\Upsilon = \mathscr{E}(-M)$. Hence:

$$X_t = x_t \int_{[0,t]} h(u)dA_u + \Upsilon_t \left(z_t - x_t \int_{[0,t]} h(u)dA^p_u\right).$$

Since condition (**A**) holds, $\mathbb{F}$-martingales are orthogonal to Υ, and by integration by parts, we derive,

$$dX_t = x_{t-}h(t)(dA_t - \Upsilon_{t-}dA_t^p) + \psi_{t-}d\Upsilon_t + \Upsilon_{t-}dz_t + \vartheta_{t-}dx_t$$

where $\vartheta_t = \int_{[0,t]} h(u)dA_u - \Upsilon_t \int_{[0,t]} h(u)dA_u^p$ and $\psi_t = z_t - x_t \int_{[0,t]} h(u)dA_u^p$. Finally, since $d\Upsilon = -\Upsilon_- dM$,

$$dX_t = \gamma_t dM_t + \sum_i \varphi_t^i dX_t^i$$

where $\gamma_t = x_{t-}h(t) - \psi_{t-}\Upsilon_{t-}, \varphi_t^i = \beta_t^i \Upsilon_{t-} + \alpha^i \vartheta_{t-}$ are $\mathbb{G}$-predictable where the $\mathbb{F}$-predictable processes (α, β) satisfying $x = x_0 + \sum_i \alpha^i \cdot X^i, z = z_0 + \sum_i \beta^i \cdot X^i$ are obtained due to the fact that x and z are $\mathbb{F}$-martingales and $(X^1, \cdots, X^n)$ has the PRP in $\mathbb{F}$.

Being true for u.i. martingales of the specific form, the result extends to local martingales using the monotone class theorem and a localization argument. □

Theorem 3.13 *Assume that X has the PRP in $\mathbb{F}$. The following two conditions are equivalent:*
(a) The filtration $\mathbb{F}$ is immersed in $\mathbb{G}$ and X_τ is $\mathscr{F}_{\tau-}$-measurable.
(b) (X, M) has the PRP in $\mathbb{G}$ and $\mathscr{G}_\tau = \mathscr{G}_{\tau-}$.

3.2.4 Change of Probability Measure

In this subsection we are interested in the impact of a change of probability measure on immersion in a progressive enlargement of filtration set-up. In the following lemma we state an auxiliary result which is followed up by Theorem 3.15 stating a stability result. We emphasize that in general, immersion between $\mathbb{F}$ and $\mathbb{G}$ is not satisfied under $\mathbb{Q}$.

Lemma 3.14 *Assume that condition* **(C)** *or* **(A)** *holds and $\mathbb{F} \hookrightarrow \mathbb{G}$. Let M be the $\mathbb{G}$-martingale given in (2.10) and γ be a $\mathbb{G}$-predictable process such that $\mathscr{E}(\gamma \cdot M)$ is a u.i. $\mathbb{G}$-martingale. Then $\mathbb{E}[\mathscr{E}(\gamma \cdot M)_t | \mathscr{F}_t] = 1$ for all $t \geq 0$.*

Proof It is enough to prove that, for any bounded $\mathscr{F}_t$-measurable r.v. ξ, one has $\mathbb{E}[\mathscr{E}(\gamma \cdot M)_t \xi] = \mathbb{E}[\xi]$. Consider the $\mathbb{F}$-martingale y given by $y_s = \mathbb{E}[\xi | \mathscr{F}_s]$. From immersion y is a $\mathbb{G}$-martingale. Condition **(C)** implies that $[y, \mathscr{E}(\gamma \cdot M)]$ is constant, since $\mathscr{E}(\gamma \cdot M)$ is a purely discontinuous $\mathbb{G}$-martingale. Similarly, condition **(A)** also implies that $[y, \mathscr{E}(\gamma \cdot M)]$ is constant, since $\mathscr{E}(\gamma \cdot M)$ is a purely discontinuous $\mathbb{G}$-martingale with a single jump at τ. Hence the process $y\mathscr{E}(\gamma \cdot M)$ is a $\mathbb{G}$-local martingale, and since the family of random variables $(y_t \mathscr{E}(\gamma \cdot M)_t,\ t \geq 0)$ is uniformly integrable, it is a $\mathbb{G}$-martingale. Therefore

$$\mathbb{E}[\xi \mathscr{E}(\gamma \cdot M)_t] = \mathbb{E}[y_t \mathscr{E}(\gamma \cdot M)_t] = \mathbb{E}[y_0] = \mathbb{E}[\xi].$$

□

Assume $\mathbb{F} \hookrightarrow \mathbb{G}$ under $\mathbb{P}$. Let Y be an $(\mathbb{F}, \mathbb{P})$-martingale and φ be an $\mathbb{F}$-predictable process s.t. $\mathscr{E}(\varphi \cdot Y)$ is a u.i. $\mathbb{G}$-martingale. Let γ be $\mathbb{G}$-predictable s.t. $\mathscr{E}(\gamma \cdot M)$ is a u.i. $\mathbb{G}$-martingale. Let $L = \mathscr{E}(\varphi \cdot Y)\mathscr{E}(\gamma \cdot M)$ and assume that L is a positive u.i. $\mathbb{G}$-martingale. Define an equivalent probability measure $\mathbb{Q}$ by

$$\frac{d\mathbb{Q}}{d\mathbb{P}} = L_\infty = \mathscr{E}(\varphi \cdot Y)_\infty \mathscr{E}(\gamma \cdot M)_\infty. \tag{3.11}$$

Theorem 3.15 *Let* $\mathbb{F} \hookrightarrow \mathbb{G}$ *under* $\mathbb{P}$ *and* $\mathbb{Q}$ *be equivalent to* $\mathbb{P}$ *given by* (3.11). *Then:*
(a) If condition **(C)** *or* **(A)** *holds, then* $\mathbb{F} \hookrightarrow \mathbb{G}$ *under* $\mathbb{Q}$.
(b) Assume that γ *in* (3.11) *is* $\mathbb{F}$*-predictable,* Z *is positive and condition* **(A)** *holds. Then, the* $\mathbb{Q}$*-Azéma supermartingale* $Z^{\mathbb{Q}}$ *of* τ *equals* $Z_t^{\mathbb{Q}} = \exp\left(\frac{1}{Z^{\mathbb{P}}}(1+\gamma) \cdot A^{p,\mathbb{P}}\right)$.

Proof Define $\ell_t := \mathbb{E}_{\mathbb{P}}[L_t|\mathscr{F}_t]$. Since φ is $\mathbb{F}$-predictable, Lemma 3.14 implies that

$$\ell_t = \mathscr{E}(\varphi \cdot Y)_t \mathbb{E}\left[\mathscr{E}(\gamma \cdot M)_t|\mathscr{F}_t\right] = \mathscr{E}(\varphi \cdot Y)_t \quad \text{and} \quad L_t = \ell_t \mathscr{E}(\gamma \cdot M)_t.$$

(a) For any bounded $(\mathbb{F}, \mathbb{Q})$-martingale X, the process $X\ell$ is an $(\mathbb{F}, \mathbb{P})$-martingale hence, from immersion under $\mathbb{P}$, a $(\mathbb{G}, \mathbb{P})$-martingale. Under condition **(C)** or condition **(A)**, $X\ell$ is orthogonal to $\mathscr{E}(\gamma \cdot M)$, hence $X\ell\mathscr{E}(\gamma \cdot M) = XL$ is a $(\mathbb{G}, \mathbb{P})$-martingale. Again by Girsanov's theorem, X is a $(\mathbb{G}, \mathbb{Q})$-martingale.
(b) By Proposition 3.9 (c), the $\mathbb{P}$-Azéma supermartingale $Z^{\mathbb{P}}$ is $Z^{\mathbb{P}} = \exp(-\Lambda^{\mathbb{P}}) := \exp(-\frac{1}{Z^{\mathbb{P}}} \cdot A^{p,\mathbb{P}})$. Hence, thanks to Bayes' formula and the form of ℓ, one has

$$\begin{aligned} Z_t^{\mathbb{Q}} &= \frac{1}{\ell_t}\mathbb{E}_{\mathbb{P}}\left[\mathbb{1}_{\{\tau>t\}}L_t|\mathscr{F}_t\right] = \mathbb{E}_{\mathbb{P}}\left[\mathbb{1}_{\{\tau>t\}} \exp\left(-\int_0^t \gamma_s \frac{1}{Z_s^{\mathbb{P}}} dA_s^{p,\mathbb{P}}\right)\Big|\mathscr{F}_t\right] \\ &= \exp\left(-\int_0^t (1+\gamma_s)\frac{1}{Z_s^{\mathbb{P}}} dA_s^{p,\mathbb{P}}\right). \end{aligned}$$

□

Remark 3.16 (a) Assume that the hypotheses of Theorem 3.15 (b) hold and that the $(\mathbb{F}, \mathbb{P})$-dual predictable projection of τ, $A^{p,\mathbb{P}}$, is absolutely continuous w.r.t. Lebesgue's measure. Then, as shown in Remark 2.21, there exists an $\mathbb{F}$-predictable process a such that $A_t^{p,\mathbb{P}} = \int_0^t a_s ds$. Theorem 3.15 (b) and Proposition 3.9 (c) imply that the $(\mathbb{F}, \mathbb{P})$-intensity rate of τ is $\lambda^{\mathbb{P}} = \frac{a}{Z^{\mathbb{P}}}$ and, as an effect of change of probability measure, the $(\mathbb{F}, \mathbb{Q})$-intensity rate of τ is $\lambda^{\mathbb{Q}} = (1+\gamma)\frac{a}{Z^{\mathbb{P}}}$.
(b) Recall the Cox construction from Sect. 2.3. If $\mathbb{P}$ is a probability such that Θ is independent of $\mathscr{F}_\infty$ and $\mathbb{Q}$ a probability equivalent to $\mathbb{P}$, in general, it is not true that Θ is independent of $\mathscr{F}_\infty$ under $\mathbb{Q}$. Changes of probability that preserve the independence of Θ and $\mathscr{F}_\infty$ may change the law of Θ, hence the intensity rate, as we have seen in the above theorem.

3.3 Multi-default Setting

3.3.1 Successive Enlargements

Proposition 3.17 *Let τ_1 and τ_2 be two random times satisfying $\tau_1 < \tau_2$ a.s. For $i = 1, 2$ let $\mathbb{A}^i$ be the filtration generated by the default process $A^i = \mathbb{1}_{[\![\tau_i,\infty[\![}$ and $\mathbb{G} = \mathbb{F} \vee \mathbb{A}^1 \vee \mathbb{A}^2$. Then, the two following assertions are equivalent:*
(a) $\mathbb{F}$ is immersed in $\mathbb{G}$,
(b) $\mathbb{F} \hookrightarrow \mathbb{G}^1 := \mathbb{F} \vee \mathbb{A}^1$ and $\mathbb{G}^1 \hookrightarrow \mathbb{G}$.

Proof If (a) holds, then $\mathbb{F} \hookrightarrow \mathbb{G}^1$ since $\mathbb{G}^1 \subset \mathbb{G}$. It remains to check is that $\mathbb{G}^1$ is immersed in $\mathbb{G}$, or equivalently that condition (3.2) holds. This is equivalent to checking that for any bounded Borel function h, and any $\mathscr{F}_\infty$-measurable bounded random variable F_∞ one has

$$\mathbb{E}\left[F_\infty h(\tau_1)\mathbb{1}_{\{\tau_2>t\}}\right] = \mathbb{E}\left[F_\infty h(\tau_1)\mathbb{P}(\tau_2 > t|\mathscr{G}_t^1)\right].$$

We split this equality in two parts. The first equality

$$\mathbb{E}\left[F_\infty h(\tau_1)\mathbb{1}_{\{\tau_1>t\}}\mathbb{1}_{\{\tau_2>t\}}\right] = \mathbb{E}\left[F_\infty h(\tau_1)\mathbb{1}_{\{\tau_1>t\}}\mathbb{P}(\tau_2 > t|\mathscr{G}_t^1)\right]$$

holds since $\mathbb{1}_{\{\tau_1>t\}}\mathbb{1}_{\{\tau_2>t\}} = \mathbb{1}_{\{\tau_1>t\}}$, therefore $\mathbb{1}_{\{\tau_1>t\}}\mathbb{P}(\tau_2 > t|\mathscr{G}_t^1) = \mathbb{1}_{\{\tau_1>t\}}$.

Since $\mathbb{F}$ is immersed in $\mathbb{G}$, one has $\mathbb{E}[F_\infty|\mathscr{G}_t] = \mathbb{E}[F_\infty|\mathscr{F}_t]$ and, since $\mathscr{F}_t \subset \mathscr{G}_t^1 \subset \mathscr{G}_t$, it follows that $\mathbb{E}[F_\infty|\mathscr{G}_t] = \mathbb{E}[F_\infty|\mathscr{G}_t^1]$. Therefore

$$\begin{aligned}
&\mathbb{E}\left[F_\infty h(\tau_1)\mathbb{1}_{\{\tau_2>t\geq\tau_1\}}\right] = \mathbb{E}\left[\mathbb{E}[F_\infty|\mathscr{G}_t]h(\tau_1)\mathbb{1}_{\{\tau_2>t\geq\tau_1\}}\right] \\
&= \mathbb{E}\left[\mathbb{E}[F_\infty|\mathscr{G}_t^1]h(\tau_1)\mathbb{1}_{\{\tau_2>t\geq\tau_1\}}\right] = \mathbb{E}\left[\mathbb{E}\left[F_\infty|\mathscr{G}_t^1\right]\mathbb{E}\left[h(\tau_1)\mathbb{1}_{\{\tau_2>t\geq\tau_1\}}|\mathscr{G}_t^1\right]\right] \\
&= \mathbb{E}\left[F_\infty\mathbb{E}\left[h(\tau_1)\mathbb{1}_{\{\tau_2>t\geq\tau_1\}}|\mathscr{G}_t^1\right]\right] = \mathbb{E}\left[F_\infty h(\tau_1)\mathbb{1}_{\{t\geq\tau_1\}}\mathbb{E}\left[\mathbb{1}_{\{\tau_2>t\}}|\mathscr{G}_t^1\right]\right],
\end{aligned}$$

which shows that (b) holds. The implication (b) $\Rightarrow$ (a) is straightforward. □

Remark 3.18 In general, if $\mathbb{F} \subset \mathbb{G} \subset \mathbb{H}$ and $\mathbb{F} \hookrightarrow \mathbb{H}$, then $\mathbb{F} \hookrightarrow \mathbb{G}$ but it does not necessarily hold that $\mathbb{G} \hookrightarrow \mathbb{H}$ (take $\mathbb{F}$ trivial!).

3.3.2 Norros' Lemma

Lemma 3.19 *Let τ_i for $i = 1, \dots, n$ be a finite random time. Define the filtration $\mathbb{G} = \mathbb{F} \vee \mathbb{A}^1 \cdots \vee \mathbb{A}^n$ and assume that:*
(a) $\mathbb{P}(\tau_i = \tau_j) = 0$, for all $i \neq j$, $i, j = 1, \dots, n$,
(b) for all $i = 1, \dots, n$ there exist continuous increasing processes Λ^i such that the $\mathbb{G}$-martingale M^i defined in (2.10) is $M_t^i = A_t^i - \Lambda_{t\wedge\tau_i}^i$.
Then, the r.v's $\Lambda_{\tau_i}^i$ are i.i.d. with unit exponential law and are independent of $\mathscr{F}_\infty$.

Proof For any $\theta_i > -1$, the bounded process $L^i_t = (1+\theta_i)^{A^i_t} e^{-\theta_i \Lambda^i_{t\wedge\tau_i}}$, which is solution of $dL^i = L^i\_\theta_i dM^i$ is a uniformly integrable $\mathbb{G}$-martingale. Moreover, martingales M^i for $i \in \{1, \ldots, n\}$ have no common jumps, and are orthogonal. Hence $\mathbb{E}\left[\prod_{i=1}^n (1+\theta_i) e^{-\theta_i \Lambda^i_{\tau_i}}\right] = 1$, which implies

$$\mathbb{E}\left[\prod_{i=1}^n e^{-\theta_i \Lambda^i_{\tau_i}}\right] = \prod_{i=1}^n (1+\theta_i)^{-1}$$

hence the independence result and the form of the law of the r.v. $\Lambda^i_{\tau_i}$. □

3.3.3 Several Defaults in a Cox Model

Lemma 3.20 *Let $\tau_i := \inf\{t : \Lambda^i_t \geq \Theta^i\}$ be a random time for i $= 1, \ldots, n$, where the random vector $(\Theta^1, \ldots, \Theta^n)$ is independent from $\mathbb{F}$ and Λ^i's are $\mathbb{F}$-adapted increasing processes. Let $\mathbb{A}^i$ be the natural filtration of A^i, where $A^i = \mathbb{1}_{[\![\tau_i,\infty[\![}$ and $\mathbb{G} = \mathbb{F} \triangledown \mathbb{A}^1 \cdots \triangledown \mathbb{A}^n$. Then $\mathbb{F} \hookrightarrow \mathbb{G}$.*

Moreover if the Θ^i's are independent, then $\mathbb{G}^k = \mathbb{F} \triangledown \mathbb{A}^1 \cdots \triangledown \mathbb{A}^{k-1} \triangledown \mathbb{A}^k$ is immersed in $\mathbb{G}$ and the $\mathbb{G}^k$-compensator of τ_k is equal to its $\mathbb{G}$-compensator. The filtration $\mathbb{F}^k := \mathbb{F} \triangledown \mathbb{A}^k$ is immersed in $\mathbb{G}$ and the $\mathbb{F}^k$-compensator of τ_k is equal to its $\mathbb{G}$-compensator.

Proof Observe that $\mathbb{G} \subset \mathbb{F} \vee \sigma(\Theta^1) \vee \cdots \vee \sigma(\Theta^n)$. Then, $\mathbb{F}$ is immersed in $\mathbb{F} \vee \sigma(\Theta^1) \vee \cdots \vee \sigma(\Theta^n)$ by the independence hypothesis and Propositions 1.22 and 1.24 (b). Under the stronger hypothesis of the second part, $\mathbb{G}^k$ is independent of the remaining Θ's and $\mathbb{G}^k$ is immersed in $\mathbb{F} \vee \sigma(\Theta^1) \vee \cdots \vee \sigma(\Theta^n)$. □

Remark 3.21 It is important to note that in the case of Lemma 3.20, if the Θ^i are not independent, the $\mathbb{G}^k$-compensator of τ_k is not equal to its $\mathbb{G}$-compensator. In other words, $\mathbb{G}^k$ is not immersed in $\mathbb{G}$ in that general setting.

3.3.4 Ordered Times

Lemma 3.22 *Let $(\tau_i, i = 1, \ldots, n)$ be a family of ordered random times and $\mathbb{A}^i$ be the natural filtration of A^i, where $A^i = \mathbb{1}_{[\![\tau_i,\infty[\![}$. Define $\mathbb{G}^k = \mathbb{F} \triangledown \mathbb{A}^1 \cdots \triangledown \mathbb{A}^k$. Then, for any k, any $\mathbb{G}^k$-martingale stopped at τ_k is a $\mathbb{G}^n$-martingale.*

Proof We prove that any $\mathbb{G}^1$-martingale stopped at τ_1 is a $\mathbb{G}^2$-martingale. The general result will follow. Let X be a $\mathbb{G}^1$-martingale stopped at τ_1, i.e., $X_t = X_{t\wedge\tau_1}$ for any t. By Lemma 2.9, for $s < t$,

$$\mathbb{E}[X_{t\wedge\tau_1}|\mathcal{G}_s^2] = 1\!\!1_{\{\tau_2\le s\}}X_{\tau_1} + 1\!\!1_{\{s<\tau_2\}}\frac{\mathbb{E}[X_{t\wedge\tau_1}1\!\!1_{\{s<\tau_2\}}|\mathcal{G}_s^1]}{\mathbb{P}(s<\tau_2|\mathcal{G}_s^1)}.$$

The numerator of the second term on the right-hand side can be rewritten as

$$\mathbb{E}[X_{t\wedge\tau_1}1\!\!1_{\{s<\tau_2\}}|\mathcal{G}_s^1] = 1\!\!1_{\{s<\tau_1\}}\mathbb{E}[X_{t\wedge\tau_1}|\mathcal{G}_s^1] + 1\!\!1_{\{\tau_1\le s\}}\mathbb{E}[X_{\tau_1}1\!\!1_{\{s<\tau_2\}}|\mathcal{G}_s^1].$$

Then, on the one hand, since $\tau_2 > \tau_1$, one has $1\!\!1_{\{s<\tau_1\}}\mathbb{P}(s<\tau_2|\mathcal{G}_s^1) = 1\!\!1_{\{s<\tau_1\}}$. Then, the $\mathbb{G}^1$-martingale property of X yields

$$1\!\!1_{\{s<\tau_1\}}\mathbb{E}[X_{t\wedge\tau_1}|\mathcal{G}_s^1] = 1\!\!1_{\{s<\tau_1\}}X_{s\wedge\tau_1}\mathbb{P}(s<\tau_2|\mathcal{G}_s^1).$$

On the other hand,

$$1\!\!1_{\{\tau_1\le s\}}\mathbb{E}[X_{\tau_1}1\!\!1_{\{s<\tau_2\}}|\mathcal{G}_s^1] = 1\!\!1_{\{\tau_1\le s\}}X_{\tau_1}\mathbb{P}(s<\tau_2|\mathcal{G}_s^1).$$

Summing up, it follows that $\mathbb{E}[X_{t\wedge\tau_1}|\mathcal{G}_s^2] = X_{s\wedge\tau_1}$, hence the result. □

Remark 3.23 The $\mathbb{G}^k$-compensator of τ_k is the increasing $\mathbb{G}^k$-predictable process Λ^k such that $M^k := 1\!\!1_{[\![\tau_k,\infty[\![} - \Lambda^k$ is a $\mathbb{G}^k$-martingale. The $\mathbb{G}^k$-martingale M^k is stopped at τ_k, since the $\mathbb{G}^k$-compensator of τ_k satisfies $\Lambda_t^k = \Lambda_{t\wedge\tau_k}^k$. It follows that the $\mathbb{G}^k$-compensator of τ_k coincides with its $\mathbb{G}^n$-compensator.

Corollary 3.24 *Assume that the τ_k are ordered and let Λ^k be the $\mathbb{G}^k$-compensator of τ_k. Then the compensator of the loss process $L := \sum_{k=1}^n 1\!\!1_{[\![\tau_k,\infty[\![}$, denoted Λ^L, is the sum of the compensators of τ_k's, i.e., $\Lambda^L = \sum_{k=1}^n \Lambda^k$.*

Proof Since $1\!\!1_{[\![\tau_k,\infty[\![} - \Lambda^k$ is a $\mathbb{G}^k$-martingale stopped at τ_k, it is a $\mathbb{G}^n$-martingale. By taking the sum, we obtain that $L - \sum_{k=1}^n \Lambda^k$ is a $\mathbb{G}^n$-martingale. □

3.4 Immersion and Finance

Let $\mathbb{F}$ be a filtration and $\mathbb{G}$ an enlargement of $\mathbb{F}$. Assume that an $\mathbb{F}$-adapted asset S is traded and that there exist a probability $\mathbb{Q}^{\mathbb{F}}$ on $\mathbb{F}$ such that S is a $(\mathbb{Q}^{\mathbb{F}}, \mathbb{F})$-martingale, and a probability $\mathbb{Q}^{\mathbb{G}}$ on $\mathbb{G}$ such that S is a $(\mathbb{Q}^{\mathbb{G}}, \mathbb{G})$-martingale. Then, S is a $(\mathbb{Q}^{\mathbb{G}}, \mathbb{F})$-martingale.

If $\mathbb{Q}^{\mathbb{F}}$ is unique, the market with a constant savings account (zero interest rate) is complete and any $(\mathbb{Q}^{\mathbb{F}}, \mathbb{F})$-martingale X can be written as $X_t = x + \int_0^t x_s dS_s$ with x being $\mathbb{F}$-predictable and is a $(\mathbb{Q}^{\mathbb{G}}, \mathbb{G})$-martingale as well. By the assumed uniqueness, $\mathbb{Q}^{\mathbb{F}}|_{\mathcal{F}_t} = \mathbb{Q}^{\mathbb{G}}|_{\mathcal{F}_t}$, and any $(\mathbb{Q}^{\mathbb{G}}, \mathbb{F})$-martingale is a $(\mathbb{Q}^{\mathbb{G}}, \mathbb{G})$-martingale: immersion holds under $\mathbb{Q}$. If $\mathbb{Q}$ is not unique, one has partial immersion: $\mathbb{F}$ martingales of the form $x + \int_0^t x_s dS_s$ (where x is $\mathbb{F}$-adapted) are $\mathbb{G}$-martingales, but it can happen that immersion does not hold in the sense that *not all* $\mathbb{F}$-martingales remain $\mathbb{G}$-martingales.

3.5 Bibliographic Notes

The first papers on immersion are Brémaud and Yor [45] (under the name hypothesis $\mathscr{H}$), Mazziotto and Szpirglas [175] and Stricker [204]. This notion was applied in a financial setting by Kusuoka [163], Elliott et al. [89], Coculescu et al. [61] and Jeanblanc and Rutkowski [131, 132] among others. The word immersion is due to Emery [90].

Immersion is linked with the causality effect which appears in the economic literature, in particular in Florens and Fougère [94], Granger [107], Gourieroux et al. [106] and Mykland [180, 181].

Immersion is identified in causal transport problem and further links to enlargement of filtration are exploited in Acciaio et al. [1].

A new formulation of immersion is given in Carmona and Lacker [53]. More precisely, it is shown in [53] that, if $\mathbb{F}$ is a natural filtration of a Brownian motion B and $\mathbb{G}$ is a progressive enlargement of $\mathbb{F}$ with τ, then $\mathbb{F} \hookrightarrow \mathbb{G}$ if and only if there exists a sequence of $\mathbb{F}$-stopping times $(\tau_n)_n$ such that (B, τ_n) converges to (B, τ) in distribution.

In Jeanblanc and Le Cam [129], the authors explain why financial market models often assume that immersion holds under an EMM: they prove that, in an incomplete market, if the price process is given as an $\mathbb{F}$-adapted process and is not modified when working in the enlarged filtration (even if its dynamics change) there exists an EMM for which immersion is satisfied.

Proposition 3.6 is presented in Jeulin and Yor [140]. One can find more results on immersion under change of probability in Bielecki et al. [38] and Coculescu et al. [61].

The equality of filtrations of β and $|B|$, presented in Example 3.5 is a deep result due to Emery and Perkins [91].

The result of Proposition 3.17 was obtained by Ehlers and Schönbucher [85], we have given here a slightly different proof.

The predictable representation property is a difficult result in general, even under immersion. For the progressive enlargement setting, see Coculescu et al. [61] for the case where immersion is satisfied and τ avoids $\mathbb{F}$-stopping times (in that case, the PRP in $\mathbb{F}$ extends to $\mathbb{G}$, adding the martingale M), and Jeanblanc and Song [135] for a more general setting. There is also an ongoing work by Di Tella and Engelbert [82] in this subject.

3.6 Exercises

Exercise 3.1 Let $\mathbb{F}^{\sigma(\tau)} = \mathbb{F} \triangledown \sigma(\tau)$ where τ is a random time and let $\mathbb{G} = \mathbb{F} \triangledown \mathbb{A}$. Find conditions on τ so that $\mathbb{G} \hookrightarrow \mathbb{F}^{\sigma(\tau)}$.

Exercise 3.2 Let $\mathbb{F} \hookrightarrow \mathbb{G}$. Prove that $\mathbb{E}[\int_0^t a_s ds|\mathscr{F}_t] = \int_0^t \mathbb{E}[a_s|\mathscr{F}_s]ds$ for any $\mathbb{G}$-adapted process a.

Exercise 3.3 Show that, if τ is $\mathscr{F}_\infty$-measurable, immersion holds between $\mathbb{F}$ and $\mathbb{G} = \mathbb{F} \triangledown \mathbb{A}$ if and only if τ is an $\mathbb{F}$-stopping time.

Exercise 3.4 Prove that if $\mathbb{A}$ and $\mathbb{F}$ are immersed in $\mathbb{G} = \mathbb{F} \triangledown \mathbb{A}$, and if any $\mathbb{F}$-martingale is continuous, then τ and $\mathscr{F}_\infty$ are independent.

Exercise 3.5 Let $\tau_i, i = 1, 2$ be two random times such that $\mathbb{P}(\tau_1 = \tau_2) = 0$ and $\mathbb{A}^i$ the filtration associated to τ_i. Prove that $\mathbb{A}^i, i = 1, 2$ are immersed in $\mathbb{A} := \mathbb{A}^1 \vee \mathbb{A}^2$ if and only if $\tau_i, i = 1, 2$ are independent.

Exercise 3.6 Let $\mathbb{F}$ be the Brownian filtration generated by B and $X = \mathbb{1}_{\{B>0\}} \cdot B$. Prove that the process X is an $\mathbb{F}$-martingale, however, $\mathbb{F}^X$ is not immersed in $\mathbb{F}$.

Exercise 3.7 Let $\mathbb{G}$ be a Brownian filtration generated by B and $\mathbb{F}$ the filtration generated by $\beta = \text{sgn}(B) \cdot B$. Prove that $\mathbb{F} \hookrightarrow \mathbb{G}$ and β has the PRP as well in $\mathbb{F}$ and in $\mathbb{G}$.

Chapter 4
Initial Enlargement

We study the initial enlargement of a reference filtration $\mathbb{F}$ with a random variable ζ with values in $\overline{\mathbb{R}} := [-\infty, \infty]$. The enlarged filtration $\mathbb{F}^{\sigma(\zeta)} := (\mathscr{F}_t^{\sigma(\zeta)})_{t \geq 0}$ is given by

$$\mathscr{F}_t^{\sigma(\zeta)} := \cap_{s>t} \left(\mathscr{F}_s \vee \sigma(\zeta)\right) ,$$

i.e., $\mathbb{F}^{\sigma(\zeta)} := \mathbb{F} \nabla \sigma(\zeta)$. We work in a rather general framework and we are interested in the hypothesis $\mathscr{H}'$ between $\mathbb{F}$ and $\mathbb{F}^{\sigma(\zeta)}$, see Definition 1.20 (b). We give conditions such that all $\mathbb{F}$-local martingales are $\mathbb{F}^{\sigma(\zeta)}$-semimartingales and, in that case, we provide their $\mathbb{F}^{\sigma(\zeta)}$-semimartingale decomposition. Initial enlargement of filtration can be interpreted in a financial setting as follows: an insider has, at time 0, some information about an event which will occur in the future, and he can use this information to make profit.

- We study Brownian bridges and Poisson bridges in an enlargement of filtration perspective, and we solve the associated insider trading problems.
- We present Yor's methodology in a setting based on the predictable representation property and we give examples.
- We present Jacod's condition of absolute continuity and its special equivalence case, and we give examples.
- We give conditions to prevent arbitrages.

4.1 Brownian and Poisson Bridges

4.1.1 Brownian Bridge

The first example of enlargement of filtration is the Brownian bridge, that is, roughly speaking, the case where, at time 0, the value of the Brownian motion at some terminal

A. Aksamit and M. Jeanblanc, *Enlargement of Filtration with Finance in View*, SpringerBriefs in Quantitative Finance,
https://doi.org/10.1007/978-3-319-41255-9_4

date T is known. One wants to understand the behaviour of the original Brownian motion B in the filtration enlarged by the σ-field generated by B_T. In what follows, without loss of generality, we take $T = 1$.

Let B be a Brownian motion, $\mathbb{F}$ its natural filtration and $\mathbb{F}^{\sigma(B_1)} := \mathbb{F} \nabla \sigma(B_1)$. The process B is no longer an $\mathbb{F}^{\sigma(B_1)}$-martingale. Indeed, by looking at the process $\left(\mathbb{E}[B_1|\mathscr{F}_t^{\sigma(B_1)}], t \geq 0\right)$ which is identically equal to B_1, one concludes that B is not an $\mathbb{F}^{\sigma(B_1)}$-martingale. However, B is an $\mathbb{F}^{\sigma(B_1)}$-semimartingale, as follows from the next proposition.

Proposition 4.1 *The process β defined as*

$$\beta_t := B_t - \int_0^{t\wedge 1} \frac{B_1 - B_s}{1-s} ds \tag{4.1}$$

is an $\mathbb{F}^{\sigma(B_1)}$-Brownian motion.

Proof Note that the integral $\int_0^{t\wedge 1} \frac{B_1 - B_s}{1-s} ds$ in (4.1) is absolutely convergent as

$$\mathbb{E}\left[\int_0^1 \frac{|B_1 - B_s|}{1-s} ds\right] = \sqrt{\frac{2}{\pi}} \int_0^1 \frac{1}{\sqrt{1-s}} ds < \infty.$$

Since $\mathscr{F}_s$ is independent of $(B_{s+h} - B_s, h \geq 0)$, and $\mathscr{F}_t \vee \sigma(B_1) = \mathscr{F}_t \vee \sigma(B_1 - B_t)$, one has for $s < t \leq 1$:

$$\mathbb{E}[B_t - B_s|\mathscr{F}_s \vee \sigma(B_1)] = \mathbb{E}[B_t - B_s|B_1 - B_s] = \frac{t-s}{1-s}(B_1 - B_s),$$

where the last equality comes from the computation of the L^2 projection of the random variable $(B_t - B_s)$ onto the space $\{\lambda(B_1 - B_s) : \lambda \in \mathbb{R}\}$. From Theorem 1.11, the equality $\mathbb{E}[B_t - B_s|\mathscr{F}_s^{\sigma(B_1)}] = \frac{t-s}{1-s}(B_1 - B_s)$ holds for the regularized filtration. On the other hand, for $s < t \leq 1$, we have

$$\begin{aligned}
\mathbb{E}\left[\int_s^t \frac{B_1 - B_u}{1-u} du \Big| \mathscr{F}_s^{\sigma(B_1)}\right] &= \int_s^t \frac{1}{1-u} \mathbb{E}\left[B_1 - B_u|\mathscr{F}_s^{\sigma(B_1)}\right] du \\
&= \int_s^t \frac{1}{1-u} \left(B_1 - B_s - \mathbb{E}\left[B_u - B_s|\mathscr{F}_s^{\sigma(B_1)}\right]\right) du \\
&= \int_s^t \frac{1}{1-u} \left(B_1 - B_s - \frac{u-s}{1-s}(B_1 - B_s)\right) du = \frac{t-s}{1-s}(B_1 - B_s).
\end{aligned}$$

It follows that $\mathbb{E}\left[\beta_t - \beta_s|\mathscr{F}_s^{\sigma(B_1)}\right] = 0$, hence β is an $\mathbb{F}^{\sigma(B_1)}$-martingale. Then, by Lévy's theorem (see [136, Theorem 1.4.1.2]), it is an $\mathbb{F}^{\sigma(B_1)}$-Brownian motion since it has continuous paths and $\langle \beta \rangle_t = \langle B \rangle_t = t$. □

Corollary 4.2 *(a) The process β given in* (4.1) *is independent of B_1.*
(b) The equality $\mathbb{F}^{\sigma(B_1)} = \mathbb{F}^{\beta} \nabla \sigma(B_1)$ holds and any square integrable $\mathbb{F}^{\sigma(B_1)}$-

martingale X has a representation $X = f(B_1) + H \bullet \beta$, where H is an $\mathbb{F}^{\sigma(B_1)}$-predictable process and f is a Borel function.

Proof (a) Note that for any t, β_t and B_1 are independent as uncorrelated and normally distributed. Then, since $\mathbb{P}(\beta_t \in A|\mathscr{F}_s^{\sigma(B_1)}) = f(\beta_s)$ where $t \geq s$, A is a Borel set and f is a Borel function, the assertion follows by induction.
(b) The equality of the two filtrations is straightforward. Since the process β is a Brownian motion in its natural filtration $\mathbb{F}^\beta$, it has the PRP. Enlarging initially the filtration $\mathbb{F}^\beta$ with a random variable B_1 which, by assertion (a), is independent of $\mathbb{F}^\beta$, by Proposition 1.23, preserves the PRP. Therefore the representation result for any square integrable $\mathbb{F}^{\sigma(B_1)}$-martingale follows. □

One has to be careful: the fact that B is an $\mathbb{F}^{\sigma(B_1)}$-semimartingale does not imply that any $\mathbb{F}$-local martingale is an $\mathbb{F}^{\sigma(B_1)}$-semimartingale. Indeed, consider a process X of the form $X := X_0 + \varphi \bullet B$ where φ satisfies $\int_0^1 \varphi_s^2 ds < \infty$. We cannot guarantee that φ is integrable w.r.t. the finite variation part of β from (4.1), i.e., we cannot guarantee that the quantity $\int_0^1 \frac{|B_1 - B_s|}{1-s} |\varphi_s| ds$ is well-defined (see Proposition 1.26 (b)). This is what Jeulin and Yor [142] have called *false friends* (fr. *faux amis*). Let us recall their result concerning the integrability problem.

Theorem 4.3 *Let X be an $\mathbb{F}$-local martingale with representation $X_t = X_0 + \int_0^t \varphi_s dB_s$ for an $\mathbb{F}$-predictable process φ satisfying $\int_0^1 \varphi_s^2 ds < \infty$ a.s. Then, the following conditions are equivalent:*

(a) the process X is an $\mathbb{F}^{\sigma(B_1)}$-semimartingale;
(b) $\int_0^1 |\varphi_s| \frac{|B_1 - B_s|}{1-s} ds < \infty$ $\mathbb{P}$-a.s.;
(c) $\int_0^1 \frac{|\varphi_s|}{\sqrt{1-s}} ds < \infty$ $\mathbb{P}$-a.s.
If these conditions are satisfied, the $\mathbb{F}^{\sigma(B_1)}$-semimartingale decomposition of X is

$$X_t = X_0 + \int_0^{t\wedge 1} \varphi_s d\beta_s + \int_0^{t\wedge 1} \varphi_s \frac{B_1 - B_s}{1-s} ds. \tag{4.2}$$

This is an example where hypothesis $\mathscr{H}'$ fails: some $\mathbb{F}$-martingales are $\mathbb{F}^{\sigma(B_1)}$-semimartingales, but not all of them. For example, for

$$\varphi(s) = (1-s)^{-\frac{1}{2}}(-\log(1-s))^{-\alpha} \mathbb{1}_{\{\frac{1}{2} < s < 1\}}$$

with $\alpha \in (\frac{1}{2}, 1]$ we have

$$\int_0^1 \varphi^2(s) ds < \infty \quad \text{and} \quad \int_0^1 \frac{\varphi(s)}{\sqrt{1-s}} ds = \infty,$$

therefore $(\int_0^t \varphi(s) dB_s, t \geq 0)$ is an $\mathbb{F}$-martingale but is not an $\mathbb{F}^{\sigma(B_1)}$-semimartingale.

Remark 4.4 The singularity of $\frac{B_1 - B_t}{1-t}$ at $t = 1$, i.e., the fact that $\frac{B_1 - B_t}{1-t}$ is not square integrable between 0 and 1 makes it impossible to use Girsanov's theorem with change of measure transforming the $(\mathbb{F}^{\sigma(B_1)}, \mathbb{P})$-semimartingale B into a $(\mathbb{F}^{\sigma(B_1)}, \mathbb{Q})$-martingale. This fact will unsurprisingly lead to arbitrage opportunities.

4.1.2 Poisson Bridge

Let N be a Poisson process with constant intensity λ and $\mathbb{F}^N$ its natural filtration. The process $(\widetilde{N}_t := N_t - \lambda t, t \geq 0)$ is an $\mathbb{F}^N$-martingale. Fix $T > 0$ and consider the initial enlargement of $\mathbb{F}^N$ with N_T, namely the filtration $\mathbb{F}^{\sigma(N_T)} := \mathbb{F}^N \vee \sigma(N_T)$.

Proposition 4.5 *The process $\widehat{N}$ defined as*

$$\widehat{N}_t := \widetilde{N}_t - \int_0^{t\wedge T} \frac{\widetilde{N}_T - \widetilde{N}_s}{T-s} ds$$

is an $\mathbb{F}^{\sigma(N_T)}$-martingale.

Proof The proof of this proposition uses the same arguments as the proof of Proposition 4.1. In the first step we note that, for $t \leq T$,

$$\widetilde{N}_t - \int_0^t \frac{\widetilde{N}_T - \widetilde{N}_s}{T-s} ds = N_t - \int_0^t \frac{N_T - N_s}{T-s} ds \,.$$

We write for $0 < s < t < T$,

$$\mathbb{E}\left[N_t - N_s \middle| \mathscr{F}_s^{\sigma(N_T)}\right] = \mathbb{E}\left[N_t - N_s | N_T - N_s\right] = \frac{t-s}{T-s}(N_T - N_s),$$

where the last equality follows from the fact that, if X and Y are two independent random variables, with Poisson laws with parameters μ and κ respectively, then

$$\mathbb{P}(X = k | X + Y = n) = \frac{n!}{k!(n-k)!}\alpha^k (1-\alpha)^{n-k}$$

where $\alpha = \dfrac{\mu}{\mu + \kappa}$. Hence, with analogous computations as in the Brownian case,

$$\mathbb{E}\left[\int_s^t \frac{N_T - N_u}{T-u} du \middle| \mathscr{F}_{s.}^{\sigma(N_T)}\right] = \frac{t-s}{T-s}(N_T - N_s) \,.$$

We deduce that $\widehat{N}$ is an $\mathbb{F}^{\sigma(N_T)}$-martingale. □

Remark 4.6 Note that any càdlàg $\mathbb{F}^N$-martingale is an $\mathbb{F}^{\sigma(N_T)}$-semimartingale since it is a càdlàg process of finite variation. Therefore we do not have *false friends* in the Poisson bridge case in contrast to the Brownian bridge case.

4.2 Insider Trading

In this section, we study a simple case of insider trading. We assume that an insider, i.e. an agent with private information, knows, at time 0, the value of the price of the underlying risky asset at time T^*. This information may be used to make a profit.

We present an optimization problem where an agent aims to maximize the expected value of an utility function evaluated at his terminal wealth. We choose the logarithmic utility function due to its tractability. We show that, indeed, the existence of an EMM in the enlarged filtration in the case of Brownian bridge may be violated.

4.2.1 *Brownian Bridge*

Let $dS_t = S_t(\mu dt + \sigma dB_t)$ where B is a Brownian motion and μ and σ are constants, be the dynamics of the price of a risky asset. The natural filtration of B is denoted $\mathbb{F}$. Assume that the riskless asset has a constant interest rate r and denote by $\theta := \frac{\mu - r}{\sigma}$ the risk premium.

The wealth at time t of an agent holding ϑ_t shares of the underlying risky asset and investing the amount $\vartheta_t^0 e^{rt}$ in the savings account is $X_t = \vartheta_t^0 e^{rt} + \vartheta_t S_t$ (here, ϑ and ϑ^0 are $\mathbb{F}$-adapted processes). The self-financing condition yields

$$dX_t = \vartheta_t^0 de^{rt} + \vartheta_t dS_t = rX_t dt + \vartheta_t(dS_t - rS_t dt) .$$

Restricting our attention to positive wealth processes, ans setting $\pi_t := \vartheta_t S_t / X_t$,

$$dX_t = rX_t dt + \sigma \pi_t X_t (dB_t + \theta dt), \quad X_0 = x > 0 .$$

Here π (an $\mathbb{F}$-adapted process) is the proportion of wealth invested in the risky asset and x is the initial wealth. Denoting by $X^{\pi,x}$ this wealth process, it follows that

$$\ln(X_t^{\pi,x}) = \ln x + \int_0^t \left(r - \frac{1}{2}\pi_s^2 \sigma^2 + \theta \pi_s \sigma \right) ds + \sigma \int_0^t \pi_s dB_s .$$

In the first step, we restrict our attention to strategies π such that the local martingale represented by the stochastic integral $\pi \cdot B$ is a martingale and we call $\mathscr{A}^m$ this set of strategies. The goal of the agent is to maximise $\mathbb{E}[\ln(X_T^{\pi,x})]$ for a given T, where

$$\mathbb{E}\left[\ln(X_T^{\pi,x})\right] = \ln x + \int_0^T \mathbb{E}\left[r - \frac{1}{2}\pi_s^2 \sigma^2 + \theta \pi_s \sigma \right] ds .$$

The optimal portfolio is $\pi_s^* = \frac{\theta}{\sigma}$ (note that $\pi^* \in \mathscr{A}^m$) and

$$\sup_{\pi \in \mathscr{A}^m} \mathbb{E}\left[\ln(X_T^{\pi,x})\right] = \ln x + T\left(r + \frac{1}{2}\theta^2\right).$$

We now check that this portfolio is optimal among all $\mathbb{F}$-adapted strategies π such that the wealth is well-defined and positive (we denote this set by $\mathscr{A}(\mathbb{F})$). For any $\pi \in \mathscr{A}(\mathbb{F})$, the concavity of the logarithm leads to

$$\mathbb{E}\left[\ln(X_T^{\pi,x}) - \ln(X_T^{\pi^*,x})\right] \leq \mathbb{E}\left[\left(X_T^{\pi,x} - X_T^{\pi^*,x}\right)\frac{1}{X_T^{\pi^*,x}}\right].$$

The process $\eta_t := x(e^{-rt}X_t^{\pi^*,x})^{-1}$ is a positive $\mathbb{P}$-martingale with initial value one: indeed $d\eta_t = -\eta_t\theta dB_t$. Denoting by $\widetilde{\mathbb{P}}$ the EMM defined on $\mathscr{F}_T$ by $d\widetilde{\mathbb{P}} = \eta_T d\mathbb{P}$, Girsanov's theorem implies that $(e^{-rt}X_t^{\pi,x}, t \geq 0)$ is a positive $\widetilde{\mathbb{P}}$-local martingale, hence a supermartingale and $\mathbb{E}_{\widetilde{\mathbb{P}}}\left[e^{-rT}X_T^{\pi,x}\right] \leq x$. Then, for any $\pi \in \mathscr{A}(\mathbb{F})$, one has $\mathbb{E}\left[\ln(X_T^{\pi,x}) - \ln(X_T^{\pi^*,x})\right] \leq 0$ and $\sup_{\pi\in\mathscr{A}(\mathbb{F})} \mathbb{E}[\ln(X_T^{\pi,x})] \leq \mathbb{E}[\ln(X_T^{\pi^*,x})]$.

We now enlarge initially the filtration $\mathbb{F}$ with S_{T^*} (or equivalently, with B_{T^*}) where $T < T^*$, obtaining $\mathbb{F}^{\sigma(B_{T^*})} := \mathbb{F}\nabla\sigma(B_{T^*})$. In $\mathbb{F}^{\sigma(B_{T^*})}$ the dynamics of S are

$$dS_t = S_t((\mu + \sigma\alpha_t)dt + \sigma d\beta_t) \quad \text{for } t \leq T,$$

where $\alpha_t = \frac{B_{T^*} - B_t}{T^* - t}$ and β, defined as in Proposition 4.1, is an $\mathbb{F}^{\sigma(B_{T^*})}$-Brownian motion (note that, since $T < T^*$, the conditions of Theorem 4.3 are satisfied) and the dynamics of the wealth are

$$dX_t^{\pi,x} = rX_t^{\pi,x}dt + \pi_t\sigma X_t^{\pi,x}(d\beta_t + \widetilde{\theta}_t dt)$$

with $\widetilde{\theta}_t := \theta + \alpha_t = \frac{\mu - r}{\sigma} + \frac{B_{T^*} - B_t}{T^* - t}$, so that, for $T < T^*$

$$\ln\left(X_T^{\pi,x}\right) = \ln x + \int_0^T \left(r - \frac{1}{2}\pi_s^2\sigma^2 + \sigma\pi_s\widetilde{\theta}_s\right)ds + \sigma\int_0^T \pi_s d\beta_s.$$

Using similar arguments as for an $\mathbb{F}$-informed agent, we obtain that the portfolio which maximizes $\mathbb{E}[\ln(X_T^{\pi,x})]$ over $\pi \in \mathscr{A}(\mathbb{F}^{\sigma(B_{T^*})})$ is $\pi_s^* = \frac{\widetilde{\theta}_s}{\sigma}$. Then,

$$\mathbb{E}\left[\ln(X_T^{\pi^*,x})\right] = \ln x + \int_0^T \mathbb{E}\left[r + \frac{1}{2}\widetilde{\theta}_s^2\right]ds = \ln x + \left(r + \frac{1}{2}\theta^2\right)T + \frac{1}{2}\int_0^T \mathbb{E}\left[\alpha_s^2\right]ds,$$

where we have used the fact that $\mathbb{E}[\alpha_s] = 0$. Let

$$V^{\mathbb{F}}(x) := \sup\left\{\mathbb{E}\left[\ln\left(X_T^{\pi,x}\right)\right] \,:\, \pi \in \mathscr{A}(\mathbb{F})\right\},$$
$$V^{\mathbb{F}^{\sigma(B_T)}}(x) := \sup\left\{\mathbb{E}\left[\ln\left(X_T^{\pi,x}\right)\right] \,:\, \pi \in \mathscr{A}(\mathbb{F}^{\sigma(B_{T^*})})\right\}.$$

Then, for $T < T^*$, one has $V^{\mathbb{F}^{\sigma(B_{T^*})}}(x) = V^{\mathbb{F}}(x) + \frac{1}{2}\mathbb{E}\left[\int_0^T \alpha_s^2 ds\right] = V^{\mathbb{F}}(x) + \frac{1}{2}\ln\frac{T^*}{T^*-T}$.

If $T = T^*$, the integral $\int_0^{T^*} \pi_s^* dB_s$ is not well-defined and the integral $\int_0^{T^*} \mathbb{E}[\alpha_s^2]ds$ diverges. The agent can optimise his portfolio till time $T^* - \varepsilon$ and then invest in the riskless asset, so that his terminal wealth satisfies

$$\mathbb{E}\left[\ln(X_{T^*})\right] = e^{r\varepsilon}\left(\ln x + \left(r + \frac{1}{2}\theta^2\right)(T^* - \varepsilon) + \frac{1}{2}\ln\frac{T^*}{\varepsilon}\right)$$

and the value function $V^{\mathbb{F}^{\sigma(B_{T^*})}}$ goes to infinity sending ε to 0. There does not exist an $\mathbb{F}^{\sigma(B_{T^*})}$-EMM. A candidate for the Radon–Nikodym density of an EMM, i.e., the processes L must satisfy

$$dL_t = \frac{\mu - r + \sigma\alpha_t}{\sigma} L_t d\beta_t, \quad L_0 \in L^1_+(\sigma(B_1)) \quad \text{and} \quad \mathbb{E}[L_0] = 1,$$

but it is not well-defined on the interval $[0, T^*]$. However, for any $\varepsilon \in (0, T^*]$, there exists a probability measure on $\mathcal{F}^{\sigma(B_{T^*})}_{T^*-\varepsilon}$ defined by $d\mathbb{Q} = L_{T^*-\varepsilon}d\mathbb{P}$, such that the process $(e^{-rt}S_t, t \le T^* - \varepsilon)$ is a $(\mathbb{F}^{\sigma(B_{T^*})}, \mathbb{Q})$-martingale.

To conclude, knowledge of the value of the underlying asset creates a profit, or even an arbitrage executed at the time T^*.

4.2.2 Poisson Bridge

Suppose that the interest rate is null and that the risky asset has dynamics

$$dS_t = S_{t-}\left(\mu dt + \sigma dB_t + \phi d\widetilde{N}_t\right),$$

where $\widetilde{N}$ is the compensated martingale of a standard Poisson process N, independent of the Brownian motion B, with intensity λ, and where μ, σ, ϕ are constants with $1 + \phi > 0$, to ensure that the price process S is positive. The natural filtration of S, denoted by $\mathbb{F}$, is equal to the natural filtration of B and N. Let $X^{x,\pi}$ be the wealth of an agent whose portfolio is described by π, where π_t, as in the previous subsection, is the proportion of wealth invested in the asset S at time t and π is assumed to be $\mathbb{F}$-predictable. Then

$$dX_t^{x,\pi} = \pi_t X_{t-}^{x,\pi}(\mu dt + \sigma dB_t + \phi d\widetilde{N}_t), \quad X_0^{x,\pi} = x > 0. \tag{4.3}$$

The solution of (4.3) is, if $(1 + \pi\phi) > 0$ (to obtain a positive wealth),

$$X_t^{x,\pi} = x \exp\left(\int_0^t \pi_s(\mu - \phi\lambda)ds + \int_0^t \sigma\pi_s dB_s - \frac{1}{2}\int_0^t \sigma^2\pi_s^2 ds\right)$$
$$\exp\left(\int_0^t \ln(1+\pi_s\phi)d\widetilde{N}_s + \int_0^t \lambda\ln(1+\pi_s\phi)ds\right).$$

Restricting our attention to π such that the stochastic integrals with respect to B and $\widetilde{N}$ are martingales and denoting this class $\mathscr{A}^m$, we obtain, for $\pi \in \mathscr{A}^m$,

$$\mathbb{E}\left[\ln(X_T^{x,\pi})\right] = \ln(x) + \int_0^T \mathbb{E}\left[\mu\pi_s - \frac{1}{2}\sigma^2\pi_s^2 + \lambda(\ln(1+\phi\pi_s) - \phi\pi_s)\right]ds.$$

Our aim is to solve

$$V(x) = \sup_{\pi\in\mathscr{A}^m} \mathbb{E}\left[\ln(X_T^{x,\pi})\right].$$

The maximum attainable wealth for the uninformed agent using $\mathbb{F}$-predictable strategies is obtained using the constant strategy $\widetilde{\pi}$:

$$\widetilde{\pi} = \frac{1}{2\sigma^2\phi}\left(\mu\phi - \phi^2\lambda - \sigma^2 \pm \sqrt{(\mu\phi - \phi^2\lambda - \sigma^2)^2 + 4\sigma^2\phi\mu}\right).$$

The sign in front of the square root to be used depends on the sign of quantities related to the parameters. The optimal $\widetilde{\pi}$ is the only one such that $1+\phi\widetilde{\pi} > 0$. Using the same methodology as in the Brownian case, it is easy to check that this portfolio is optimal among all the portfolios which lead to positive wealth.

We assume now that the informed agent knows N_T from time 0, and we introduce $\mathbb{F}^{\sigma(N_T)} = \mathbb{F}\vee\sigma(N_T)$. Therefore, in $\mathbb{F}^{\sigma(N_T)}$ the wealth has the dynamics

$$dX_t = \pi_t X_{t-}\left((\mu + \phi\kappa_t)dt + \sigma dB_t + \phi d\widehat{N}_t\right),$$

where $\widehat{N}$ is given in Proposition 4.5 and $\kappa_t = \dfrac{N_T - N_t}{T-t}$. Note that, by the independence hypothesis, B is an $\mathbb{F}^{\sigma(N_T)}$-Brownian motion. Hence, exactly the same computations as above can be carried out, it only requires changing μ to $\mu + \phi\kappa_t$ and the intensity of the jumps from λ to κ_t. In particular

$$X_t = x \exp\left(\int_0^t \pi_s\mu ds + \int_0^t \ln(1+\pi_s\phi)d\widehat{N}_s + \int_0^t \kappa_s\ln(1+\pi_s\phi)ds\right)\mathscr{E}(\sigma\pi\cdot B)_t.$$

The optimal portfolio π^* is given by

$$\pi_s^* = \frac{1}{2\sigma^2\phi}\left(\mu\phi - \sigma^2 \pm \sqrt{(\mu\phi + \sigma^2)^2 + 4\sigma^2\phi^2\kappa_s}\right).$$

The optimal portfolio of the uninformed agent is constant, while the optimal portfolio of the informed agent is time-varying and has a jump whenever the price process jumps.

Since, as expected, $\sup_{\pi \in \mathscr{A}(\mathbb{F}^{\sigma(N_T)})} \mathbb{E}[\ln X_T^\pi] \geq \sup_{\pi \in \mathscr{A}(\mathbb{F})} \mathbb{E}[\ln X_T^\pi]$, the maximum expected wealth for the informed agent is greater than that of the uninformed agent. In this setting, NFLVR($\mathbb{F}^{\sigma(N_T)}$) holds: indeed, in this incomplete market, the set of $\mathbb{F}^{\sigma(N_T)}$-EMM's is characterized by the set of the Radon–Nikodym densities which are positive martingales of the form

$$dL_t = L_{t-}(\psi_t dB_t + \gamma_t d\widehat{N}_t), \quad L_0 = 1 \tag{4.4}$$

where ψ and γ are predictable processes satisfying $1 + \gamma > 0$,

$$\mu + \phi\kappa_t + \sigma\psi_t + \phi\kappa_t\gamma_t = 0\ , \quad d\mathbb{P} \otimes dt\ a.s.,$$

and some integrability condition to ensure that L is a martingale (e.g., γ constant).

4.2.3 Information Drift in an Enlargement of Filtration Setting

Let B be a Brownian motion and $\mathbb{F}$ its natural filtration representing knowledge of a regular agent. We study a financial market where a risky asset with price S and a riskless asset $S^0 \equiv 1$ are traded in an arbitrage free manner in $\mathbb{F}$. We assume w.l.o.g. that S is a positive $\mathbb{F}$-(local) martingale given by $dS_t = S_t\sigma_t dB_t$. We also assume that the informed agent has access to a filtration $\mathbb{G}$, larger than $\mathbb{F}$, and that any $\mathbb{F}$-martingale is a $\mathbb{G}$-semimartingale. More precisely, we assume that there exists an integrable $\mathbb{G}$-adapted process $\mu^{\mathbb{G}}$ such that $dB_t = dB_t^{\mathbb{G}} + \mu_t^{\mathbb{G}} dt$ where $B^{\mathbb{G}}$ is a $\mathbb{G}$-Brownian motion and $\mathbb{E}\left[\int_0^T (\mu_s^{\mathbb{G}})^2 ds\right] < \infty$. The process $\mu^{\mathbb{G}}$ is called the information drift. We emphasize that $\mathbb{G}$ is a general filtration, and it is not assumed to be an initial or progressive enlargement of $\mathbb{F}$.

Let $X^{\pi,x}$ be the positive wealth process associated with an $\mathbb{F}$-predictable strategy π and an initial wealth $x > 0$

$$dX_t^{\pi,x} = \sigma_t \pi_t X_t^{\pi,x} dB_t\ , \quad X_0^{\pi,x} = x.$$

Then, using the fact that $X^{\pi,x}$ is an $\mathbb{F}$-supermartingale (as a positive local martingale), the concavity of the logarithm implies that $\sup_{\pi \in \mathscr{A}(\mathbb{F})} \mathbb{E}[\ln X_T^{\pi,x}] = \ln(x)$, where $\mathscr{A}(\mathbb{F})$ is the set of $\mathbb{F}$-adapted processes such that $X^{\pi,x}$ is positive, and the optimal π is $\pi^{*,\mathbb{F}} \equiv 0$, and $X^{\pi^*,x} \equiv x$.

Let us now consider the case of $\mathbb{G}$-predictable strategies:

$$X_t^{\pi,x} = x \exp\left(\int_0^t \sigma_s \pi_s dB_s^{\mathbb{G}} - \frac{1}{2} \int_0^t \sigma_s^2 \pi_s^2 ds + \int_0^t \sigma_s \pi_s \mu_s^{\mathbb{G}} ds \right).$$

The optimal $\mathbb{G}$-predictable π is $\pi^{*,\mathbb{G}} = \mu^{\mathbb{G}}/\sigma$ and

$$\ln X_t^{\pi^{*,\mathbb{G}},x} = \ln x + \int_0^t \mu_s^{\mathbb{G}} dB_s^{\mathbb{G}} + \frac{1}{2} \int_0^t (\mu_s^{\mathbb{G}})^2 ds$$

so that,

$$\sup_{\pi \in \mathscr{A}(\mathbb{F})} \mathbb{E}\left[\ln X_T^{\pi,x}\right] = \ln x \le \sup_{\pi \in \mathscr{A}(\mathbb{G})} \mathbb{E}\left[\ln X_T^{\pi,x}\right] = \ln x + \frac{1}{2}\mathbb{E}\left[\int_0^T (\mu_s^{\mathbb{G}})^2 ds\right].$$

Note that, in the case of the Brownian bridge, $\mathbb{E}\left[\int_0^T (\mu_s^{\mathbb{G}})^2 ds\right] = \infty$, this case being associated with arbitrage opportunities.

The case where S is a semimartingale of the form $dS_t = S_t(\alpha_t dt + \sigma_t dB_t)$ can be studied with the same methodology, assuming some integrability conditions on the coefficients.

Proposition 4.7 *Let S be a positive $\mathbb{F}$-semimartingale satisfying NUPBR and U be a concave utility function defined on $\mathbb{R}^+$. Assume*[1] *that, for $x > 0$, the utility maximization problem $\max_{\pi \in \mathscr{A}(\mathbb{F})} \mathbb{E}\left[U(X_T^{\pi,x})\right]$ has a solution $X_T^* = (U')^{-1}(\lambda L_T^*)$ where L^* is an $\mathbb{F}$-local martingale deflator for S such that $L_0^* = 1$ and $\lambda \in \mathbb{R}$. If $\mathbb{F} \hookrightarrow \mathbb{G}$, then*

$$\max_{\pi \in \mathscr{A}(\mathbb{F})} \mathbb{E}\left[U(X_T^{\pi,x})\right] = \max_{\pi \in \mathscr{A}(\mathbb{G})} \mathbb{E}\left[U(X_T^{\pi,x})\right]$$

where $\mathscr{A}(\mathbb{F})$ (resp. $\mathscr{A}(\mathbb{G})$) is the set of $\mathbb{F}$-predictable (resp. $\mathbb{G}$ predictable) processes π such that $X^{\pi,x}$ is positive.

Proof Being an $\mathbb{F}$-local martingale deflator for S, by immersion, L^* is a $\mathbb{G}$-local martingale deflator for S. Hence for any strategy $\pi \in \mathscr{A}(\mathbb{G})$, the process $X^{\pi,x}L^*$ is a $\mathbb{G}$-supermartingale (if one restricts attention to strategies such that the wealth is non-negative) with initial value x. By concavity of the utility function

$$\mathbb{E}\left[U(X_T^{\pi,x}) - U(X_T^*)\right] \le \mathbb{E}\left[(X_T^{\pi,x} - X_T^*)U'(X_T^*)\right] = \lambda \mathbb{E}\left[(X_T^{\pi,x} - X_T^*)L_T^*\right] \le 0$$

which proves that X_T^* is optimal for $\mathscr{A}(\mathbb{G})$-strategies. □

Remark 4.8 Note that, if $\mathscr{E}(-\mu^{\mathbb{G}} \bullet B^{\mathbb{G}})$ is a $\mathbb{G}$-martingale, NFLVR($\mathbb{G}$) holds, and if it is a $\mathbb{G}$-local martingale, NUPBR($\mathbb{G}$) holds (see Sect. 1.4).

[1] Under mild conditions on the semimartingale S and on the utility function U, this assumption is satisfied, see [161].

4.3 Yor's Method

In this section we present an approach to solve the problem of initial enlargement of the natural filtration $\mathbb{F}$ of a Brownian motion B with a $\overline{\mathbb{R}}$-valued r.v. ζ. In this case, the quadratic covariation process of $\mathbb{F}$-local martingales does not depend on the filtration. The method strongly relies on the fact that B has the predictable representation property in $\mathbb{F}$.

For a bounded Borel function $f : \overline{\mathbb{R}} \to \mathbb{R}$, let $(\lambda_t(f), t \geq 0)$ be the continuous modification of the martingale $(\mathbb{E}[f(\zeta)|\mathcal{F}_t], t \geq 0)$. Relying on Riesz's theorem on the representation of linear functionals, one can prove that there exists a σ-finite predictable kernel $\lambda_t(dx)$ such that

$$\lambda_t(f) = \int_{\overline{\mathbb{R}}} f(x)\lambda_t(dx) .$$

From the predictable representation property applied to the martingale $\mathbb{E}[f(\zeta)|\mathcal{F}_t]$, there exists a predictable process $\widehat{\lambda}(f)$ such that

$$\lambda_t(f) = \mathbb{E}[f(\zeta)] + \int_0^t \widehat{\lambda}_s(f)dB_s .$$

Proposition 4.9 *Assume that there exists a predictable kernel $\widehat{\lambda}_t(dx)$ such that*

$$\widehat{\lambda}_t(f) = \int_{\overline{\mathbb{R}}} f(x)\ \widehat{\lambda}_t(dx) \quad d\mathbb{P} \otimes dt \ a.s.$$

Assume furthermore that $d\mathbb{P} \otimes dt$ a.s. the measure $\widehat{\lambda}_t(dx)$ is absolutely continuous with respect to $\lambda_t(dx)$ and define $\rho(x,t)$ by

$$\widehat{\lambda}_t(dx) = \rho(x,t)\lambda_t(dx) .$$

Then, for any $\mathbb{F}$-local martingale X satisfying $\int_0^t |\rho(\zeta,s)||d\langle X,B\rangle_s|$, there exists an $\mathbb{F}^{\sigma(\zeta)}$-local martingale $\widetilde{X}$ such that

$$X_t = \widetilde{X}_t + \int_0^t \rho(\zeta,s)d\langle X,B\rangle_s. \tag{4.5}$$

Proof By localization it is enough to consider bounded X. Let X be a bounded $\mathbb{F}$-martingale, f a given bounded Borel function and $Y_t := \lambda_t(f) = \mathbb{E}[f(\zeta)|\mathcal{F}_t]$. Thus the processes $X \bullet Y$ and $Y \bullet X$ are $\mathbb{F}$-martingales. Then, for $A_s \in \mathcal{F}_s$ and $s < t$:

$$\mathbb{E}[\mathbb{1}_{A_s} f(\zeta)(X_t - X_s)] = \mathbb{E}[\mathbb{1}_{A_s}(Y_t X_t - Y_s X_s)] = \mathbb{E}[\mathbb{1}_{A_s}(\langle Y, X\rangle_t - \langle Y, X\rangle_s)]$$
$$= \mathbb{E}\left[\mathbb{1}_{A_s}\int_s^t \widehat{\lambda}_u(f)d\langle X, B\rangle_u\right] = \mathbb{E}\left[\mathbb{1}_{A_s}\int_s^t\int_{\mathbb{R}} f(x)\rho(x,u)\lambda_u(dx)d\langle X, B\rangle_u\right].$$

Therefore, by Fubini's theorem, $V_t := \int_0^t \rho(\zeta, u)\, d\langle X, B\rangle_u$ satisfies

$$\mathbb{E}[\mathbb{1}_{A_s} f(\zeta)(X_t - X_s)] = \mathbb{E}[\mathbb{1}_{A_s} f(\zeta)(V_t - V_s)].$$

By the monotone class theorem it follows that, for any $G_s \in \mathscr{F}_s \vee \sigma(\zeta)$,

$$\mathbb{E}[\mathbb{1}_{G_s}(X_t - X_s)] = \mathbb{E}[\mathbb{1}_{G_s}(V_t - V_s)],$$

hence, $X - V$ is an $\mathbb{F} \vee \sigma(\zeta)$-martingale and by Theorem 1.9, an $\mathbb{F}^{\sigma(\zeta)}$-martingale. □

We now give some examples taken from Mansuy and Yor [173].

Example 4.10 (*Enlargement with* B_1) Let $\zeta = B_1$, i.e., we are back to the Brownian bridge case presented in Sect. 4.2.1. From the Markov property, for any bounded Borel function g,

$$\mathbb{E}[g(B_1)|\mathscr{F}_t] = \mathbb{E}[g(B_1 - B_t + B_t)|\mathscr{F}_t] = F_g(B_t, 1-t)$$

where $F_g(y,s) = \int g(x)P(s;y,x)dx$ and $P(s;y,x) = \frac{1}{\sqrt{2\pi s}}\exp\left(-\frac{(x-y)^2}{2s}\right)$.

It follows that $\lambda_t(dx) = \frac{1}{\sqrt{2\pi(1-t)}}\exp\left(-\frac{(x-B_t)^2}{2(1-t)}\right)dx$. It remains to apply Itô's lemma to $\exp\left(-\frac{(x-B_t)^2}{2(1-t)}\right)$, and we obtain

$$\widehat{\lambda}_t(dx) = \frac{x - B_t}{1-t}\frac{1}{\sqrt{2\pi(1-t)}}\exp\left(-\frac{(x-B_t)^2}{2(1-t)}\right)dx,$$

and $\rho(x,t) = \frac{x-B_t}{1-t}$. We recover formula (4.1).

Example 4.11 (*Enlargement with* $\sup_{s\le 1} B_s$) Let $B_t^* = \sup_{s\le t} B_s$ and $\zeta = B_1^*$. From [136, Exercise 3.1.6.7], for any bounded Borel function f,

$$\mathbb{E}[f(B_1^*)|\mathscr{F}_t] = F(1-t, B_t, B_t^*)$$

where

$$F(s,a,b) = \sqrt{\frac{2}{\pi s}}\left(f(b)\int_0^{b-a} e^{-u^2/(2s)}du + \int_b^\infty f(u)e^{-(u-a)^2/(2s)}du\right)$$

and, denoting by δ_a the Dirac measure at a, $\lambda_t(dx)$ equals

$$\sqrt{\frac{2}{\pi(1-t)}}\left\{\delta_{B_t^*}(dx)\int_0^{B_t^*-B_t}\exp\left(-\frac{u^2}{2(1-t)}\right)du+\mathbb{1}_{\{x>B_t^*\}}\exp\left(-\frac{(x-B_t)^2}{2(1-t)}\right)dx\right\}.$$

Hence, by applying Itô's formula, $\widehat{\lambda}_t(dx)$ is equal to

$$\sqrt{\frac{2}{\pi(1-t)}}\left\{-\delta_{B_t^*}(dx)\exp\left(-\frac{(B_t^*-B_t)^2}{2(1-t)}\right)+\mathbb{1}_{\{x>B_t^*\}}\frac{x-B_t}{1-t}\exp\left(-\frac{(x-B_t)^2}{2(1-t)}\right)dx\right\}.$$

It follows that, setting $\varphi(x)=\exp(-x^2/2)/\int_0^x\exp(-u^2/2)du$,

$$\rho(x,t)=\mathbb{1}_{\{x>B_t^*\}}\frac{x-B_t}{1-t}-\mathbb{1}_{\{B_t^*=x\}}\frac{1}{\sqrt{1-t}}\varphi\left(\frac{B_t^*-B_t}{\sqrt{1-t}}\right)$$

and that, for an $\mathbb{F}$-martingale X,

$$X_t=\widetilde{X}_t+\int_0^{t\wedge\tau}\frac{B_1^*-B_s}{1-s}d\langle X,B\rangle_s-\int_\tau^t\frac{1}{\sqrt{1-s}}\varphi\left(\frac{B_s^*-B_s}{\sqrt{1-s}}\right)d\langle X,B\rangle_s$$

where $\tau=\inf\{t\,:\,B_t=B_1^*\}$ and $\widetilde{X}$ is an $\mathbb{F}^{\sigma(B_1^*)}$-martingale.

Example 4.12 (Enlargement with the Supremum of a Martingale) Let Y be a non-negative local martingale in a Brownian filtration $\mathbb{F}$, such that $Y_0=1$ and $\lim_{t\to\infty}Y_t=0$. We write $Y_t^*:=\sup\{Y_s,\,s\le t\}$. We initially enlarge the filtration $\mathbb{F}$ with Y_∞^*. For any bounded Borel function f, introducing $Y^{*,t}:=\sup_{s\ge t}Y_s$, one has

$$\begin{aligned}\lambda_t(f):=\mathbb{E}[f(Y_\infty^*)|\mathscr{F}_t]&=\mathbb{E}[f(Y_t^*\vee Y^{*,t})|\mathscr{F}_t]\\&=f(Y_t^*)\mathbb{P}(Y_t^*>Y^{*,t}|\mathscr{F}_t)+\mathbb{E}[f(Y^{*,t})\mathbb{1}_{\{Y^{*,t}>Y_t^*\}}|\mathscr{F}_t].\end{aligned}$$

From Doob's maximal lemma (see Exercise 1.6), conditionally on $\mathscr{F}_t$, the random variable $Y^{*,t}$ is distributed as Y_t/U where U is a uniform random variable, independent of $\mathscr{F}_t$. It follows that

$$\lambda_t(f)=f(Y_t^*)\left(1-\frac{Y_t}{Y_t^*}\right)+\int_0^{Y_t/Y_t^*}f\left(\frac{Y_t}{u}\right)du,$$

therefore

$$\begin{aligned}\lambda_t(dx)&=(1-(Y_t/Y_t^*))\delta_{Y_t^*}(dx)+Y_t\mathbb{1}_{[Y_t^*,\infty)}(x)\frac{dx}{x^2},\\\widehat{\lambda}_t(dx)&=-(1/Y_t^*)\delta_{Y_t^*}(dx)+\mathbb{1}_{[Y_t^*,\infty)}(x)\frac{dx}{x^2},\\\rho(x,t)&=1/(Y_t-Y_t^*)\mathbb{1}_{\{Y_t^*=x\}}+\frac{1}{Y_t}\mathbb{1}_{\{Y_t^*<x\}},\end{aligned}$$

which leads to the following decomposition of an $\mathbb{F}$-martingale X,

$$X_t = \widetilde{X}_t + \int_0^{t\wedge\tau} \frac{d\langle X, Y\rangle_s}{Y_s} + \int_\tau^{\tau\vee t} \frac{d\langle X, Y\rangle_s}{Y_s - Y_s^*}, \tag{4.6}$$

where $\widetilde{X}$ is an $\mathbb{F}^{\sigma(Y_\infty^*)}$-local martingale and

$$\tau := \sup\{t \geq 0 \,:\, Y_t^* = Y_t\} = \sup\{t \geq 0 \,:\, Y_t = Y_\infty^*\}.$$

4.4 Jacod's Absolute Continuity Condition

In this section we study another generic approach to the initial enlargement of filtration problem. In comparison to Yor's method, it does not rely on the predictable representation property. Instead, it is based on an assumption on the conditional law of the $\overline{\mathbb{R}}$-valued random variable ζ.

4.4.1 Generalities

In order to formulate Jacod's condition, let us recall that a family of $\mathbb{F}$-regular conditional distributions of ζ given $\mathbb{F}$, denoted $P_t(\omega, dx)$, is a family of probability measures indexed by t and $\omega \in \Omega$ such that, for every t and every $A \in \mathscr{B}(\overline{\mathbb{R}})$, $P_t(\cdot, A)$ is a version of $\mathbb{P}(\zeta \in A|\mathscr{F}_t)$. Since ζ is $\overline{\mathbb{R}}$-valued, such a family always exists.

Definition 4.13 Let η be the law of ζ. We say that ζ satisfies **Jacod's absolute continuity condition** if, for each $t \geq 0$, its conditional law is absolutely continuous with respect to its unconditional law, i.e.,

$$P_t(du) \ll \eta(du), \quad a.s.$$

We say that ζ satisfies **Jacod's equivalence condition** if, for each $t \geq 0$, its conditional law is equivalent to its unconditional law, i.e.,

$$P_t(du) \sim \eta(du), \quad a.s.$$

Remark 4.14 (a) Note that, in the Brownian bridge case (i.e., for $\zeta = B_1$), Jacod's absolute continuity condition does not hold (it fails at time 1).
(b) Note that Jacod's conditions are invariant with respect to a change of probability measure equivalent on $\mathscr{F}_t^{\sigma(\zeta)}$ for each $t \geq 0$. Indeed if $\mathbb{Q}$ is a probability measure on $(\Omega, \mathscr{F})$ which is equivalent to $\mathbb{P}$ on $\mathscr{F}_t^{\sigma(\zeta)}$ for each $t \geq 0$ then, by the fact that ζ is $\mathscr{F}_t^{\sigma(\zeta)}$-measurable, for $A \in \mathscr{B}(\overline{\mathbb{R}})$, it follows that $\mathbb{P}(\zeta \in A|\mathscr{F}_t) = 0$ $\mathbb{P}$-a.s. if and only if $\mathbb{Q}(\zeta \in A|\mathscr{F}_t) = 0$ $\mathbb{Q}$-a.s. for each $t \geq 0$. •

Example 4.15 Let ζ be a random variable taking only a countable number of values $(c_k)_{k\geq 1}$. Then, for any filtration $\mathbb{F}$, Jacod's absolute continuity condition holds. Indeed, the law of ζ is of the form

$$\eta\,(du) = \sum_{k=1}^{\infty} \mathbb{P}\,(\zeta = c_k)\,\delta_{c_k}\,(du)\,,$$

where δ_c is the Dirac measure at c, and $\mathbb{P}\,(\zeta = c_k) \neq 0$. Then $P_t\,(\omega, du)$ is absolutely continuous with respect to η, i.e., $P_t(du) = p_t(u)\eta(du)$ with Radon–Nikodym density:

$$p_t(u) = \sum_{k=1}^{\infty} \frac{\mathbb{P}\,(\zeta = c_k|\mathscr{F}_t)}{\mathbb{P}\,(\zeta = c_k)}\,\mathbb{1}_{\{u=c_k\}}.$$

Example 4.16 Let B be a Brownian motion and $\zeta = \int_0^\infty f(s)dB_s$ where f is a continuous deterministic function such that $\int_0^\infty f^2(s)ds < \infty$ and $\int_t^\infty f^2(s)ds \neq 0$ for each t. It is easy to compute $P_t\,(du)$, since conditionally on $\mathscr{F}_t$, the random variable ζ is Gaussian, with mean $x_t := \int_0^t f\,(s)\,dB_s$ and variance $\sigma^2(t) := \int_t^\infty f^2\,(s)\,ds$, namely

$$P_t(du) = \frac{1}{\sqrt{2\pi}\sigma(t)}\exp\left(-\frac{(u-x_t)^2}{2\sigma^2(t)}\right)\mathbb{1}_{\{u\in\mathbb{R}\}}.$$

We deduce that $P_t\,(du)$ is equivalent to the law $P_0(du)$ of ζ (a centered Gaussian law with variance $\sigma^2(0)$), i.e., $P_t(du) = p_t(u)P_0(du)$ with Radon–Nikodym density:

$$p_t(u) = \frac{\sigma\,(0)}{\sigma\,(t)}\exp\left(\frac{u^2}{2\sigma^2(0)} - \frac{(u-x_t)^2}{2\sigma^2(t)}\right)\mathbb{1}_{\{u\in\mathbb{R}\}}.$$

Proposition 4.17 *Let ζ be a random variable satisfying Jacod's absolute continuity condition. Then there exists a non-negative $\mathscr{O}(\mathbb{F}) \otimes \mathscr{B}(\overline{\mathbb{R}})$-measurable function $(\omega, t, u) \to p_t(\omega, u)$ càdlàg in t such that*
(a) for every $u \in \overline{\mathbb{R}}$, the process $(p_t(u), t \geq 0)$ is an $\mathbb{F}$-martingale, and if

$$R(u) := \inf\{t : p_{t-}(u) = 0\} \tag{4.7}$$

one has $p(u) > 0$ and $p_-(u) > 0$ on $[\![0, R(u)[\![$ and $p(u) = 0$ on $[\![R(u), \infty[\![$,
(b) for every $t \geq 0$, the measure $p_t(u)\eta(du)$ equals $\mathbb{P}(\zeta \in du|\mathscr{F}_t)$, in other words $\mathbb{E}[f(\zeta)|\mathscr{F}_t] = \int_{\mathbb{R}} f(u)p_t(u)\eta(du)$.

Proof Denote by $\mathbb{Q}^+$ the set of non-negative rational numbers. Let $t \in \mathbb{Q}^+$ and consider a non-negative $\mathscr{F}_t \otimes \mathscr{B}(\overline{\mathbb{R}})$-measurable function $(\omega, u) \to \widehat{p_t}(\omega, u)$ such that $P_t(du) = \widehat{p_t}(u)\eta(du)$. For $s \in [0, t] \cap \mathbb{Q}^+$, we denote by $p_{s,t}(\omega, u)$ an $\mathscr{F}_s \otimes \mathscr{B}(\overline{\mathbb{R}})$-measurable function equal to $\mathbb{E}[\widehat{p_t}(u)|\mathscr{F}_s]$ $\mathbb{P}$-a.s. for any $u \in \overline{\mathbb{R}}$ (see Lemma 3 in [205] for existence). Then, for an integrable $\mathscr{F}_s$-measurable r.v. X and a bounded Borel function g, we have, using the fact that $\int_{\overline{\mathbb{R}}} g(u)P_s(du) = \mathbb{E}[g(\zeta)|\mathscr{F}_s]$,

$$\mathbb{E}\left[X\int_{\overline{\mathbb{R}}} g(u)P_s(du)\right] = \mathbb{E}\left[X\mathbb{E}[g(\zeta)|\mathscr{F}_s]\right] = \mathbb{E}\left[X\mathbb{E}[g(\zeta)|\mathscr{F}_t]\right]$$
$$= \mathbb{E}\left[X\int_{\overline{\mathbb{R}}} g(u)\widehat{p}_t(u)\eta(du)\right] = \int_{\overline{\mathbb{R}}} g(u)\mathbb{E}[X\widehat{p}_t(u)]\eta(du)$$
$$= \int_{\overline{\mathbb{R}}} g(u)\mathbb{E}[Xp_{s,t}(u)]\eta(du) = \mathbb{E}\left[X\int_{\overline{\mathbb{R}}} g(u)p_{s,t}(u)\eta(du)\right],$$

which implies that $P_s(\omega, du)$ equals $p_{s,t}(\omega, u)\eta(du)$ for $\mathbb{P}\otimes\eta$-a.a. (ω, u). So the set $\{(\omega, u) : \widehat{p}_s(\omega, u) \neq p_{s,t}(\omega, u)\}$ is $\mathbb{P}\otimes\eta$-negligible. From Fubini's theorem, we deduce that the set

$$B = \{u \in \overline{\mathbb{R}} : \widehat{p}_s(u) = \mathbb{E}[\widehat{p}_t(u)|\mathscr{F}_s] \quad \mathbb{P}\text{-a.s. for each rational number } s \leq t\}$$

has a complement of η-measure zero. Thus, by defining $\widetilde{p}_t(u) := \widehat{p}_t(u)$ for $u \in B$ and $\widetilde{p}_t(u) := 1$ for $u \notin B$, we may suppose that

$$(\widetilde{p}_t(u))_{t\in\mathbb{Q}^+} \text{ is an } \mathbb{F}\text{-martingale for all } u \in \overline{\mathbb{R}}. \tag{4.8}$$

Let C_t be the set defined as

$$C_t := \{(\omega, u) : \lim_{r\in\mathbb{Q}^+, r\searrow s} \widetilde{p}_r(u) < \infty \text{ and } \lim_{r\in\mathbb{Q}^+, r\nearrow s} \widetilde{p}_r(u) < \infty \ \forall s \in [0, t]\}.$$

Then, $C_t \in \bigcap_{s>t}\mathscr{F}_s \otimes \mathscr{B}(\overline{\mathbb{R}})$ and for $C_t^u := \{\omega : (\omega, u) \in C_t\}$ property (4.8) and Theorem 2.43 in [112] imply that $\mathbb{P}(C_t^u) = 1$ for each $u \in \mathbb{R}$ and $t > 0$. We then set

$$p_t(\omega, u) = \begin{cases} \lim_{s\in Q^+, s\searrow t} \widetilde{p}_s(\omega, u) & \text{if } (\omega, u) \in \bigcap_{s>t} C_s \\ 1 & \text{otherwise.} \end{cases}$$

Since $p(u)$ is càdlàg and $(\omega, u) \to p_t(\omega, u)$ is $\bigcap_{s>t}\mathscr{F}_s \otimes \mathscr{B}(\overline{\mathbb{R}})$-measurable, $p(u)$ is $\mathbb{F}$-optional for any $u \in \overline{\mathbb{R}}$ and $(\omega, t, u) \to p_t(\omega, u)$ is $\mathscr{F} \otimes \mathscr{B}(\mathbb{R}^+) \otimes \mathscr{B}(\overline{\mathbb{R}})$-measurable. Hence, by Proposition 3 in [205], $(\omega, t, u) \to p_t(\omega, u)$ is $\mathscr{O}(\mathbb{F})\otimes\mathscr{B}(\overline{\mathbb{R}})$-measurable. Since $p(u)$ is a càdlàg regularization of a martingale $(\widetilde{p}_t(u))_{t\in\mathbb{Q}^+}$ and $\mathbb{F}$ is right-continuous, Theorem 2.44 in [112] implies that $(p_t(u))_{t\in\mathbb{R}^+}$ is an $\mathbb{F}$-martingale. Finally, properties of càdlàg non-negative martingales, by Theorem 2.62 in [112], imply that $p(u) > 0$ and $p_-(u) > 0$ on $[\![0, R(u)[\![$ and $p(u) = 0$ on $[\![R(u), \infty[\![$. □

In the following lemma, we compute the $\mathbb{F}$-predictable and $\mathbb{F}$-optional projections of some particular $\mathbb{F}^{\sigma(\zeta)}$-adapted processes.

Proposition 4.18 *(a) Let the function $(\omega, t, u) \to Y_t(\omega, u)$ be $\mathscr{P}(\mathbb{F})\otimes\mathscr{B}(\overline{\mathbb{R}})$-measurable, and either non-negative or bounded. Then, the $\mathbb{F}$-predictable projection of the process $(Y_t(\zeta))_{t\geq 0}$ is given by*

$$^{p,\mathbb{F}}(Y(\zeta))_t = \int_{\overline{\mathbb{R}}} Y_t(u)p_{t-}(u)\eta(du), \quad t \geq 0.$$

(b) Let the function $(\omega, t, u) \to Y_t(\omega, u)$ be $\mathscr{O}(\mathbb{F}) \otimes \mathscr{B}(\overline{\mathbb{R}})$-measurable, and either non-negative or bounded. Then, the $\mathbb{F}$-optional projection of the process $(Y_t(\zeta))_{t\geq 0}$ is given by

$$^{o,\mathbb{F}}(Y(\zeta))_t = \int_{\overline{\mathbb{R}}} Y_t(u)p_t(u)\eta(du), \quad t \geq 0. \tag{4.9}$$

Proof (a) It is enough to consider processes of the form $Y_t(\omega, u) = x_t(\omega)g(u)$ where x is a bounded $\mathbb{F}$-predictable process and g is a bounded $\mathscr{B}(\overline{\mathbb{R}})$-measurable function, and then argue by the monotone class theorem. In this simple case, the result follows from Proposition 4.17.

The proof of (b) is analogous. □

Proposition 4.18 implies in particular that

$$R(\zeta) = \infty \quad \mathbb{P}\text{-a.s.} \tag{4.10}$$

where $R(u)$ is defined in (4.7), or equivalently $p_t(\zeta) > 0$ and $p_{t-}(\zeta) > 0$ for $t \geq 0$ $\mathbb{P}$-a.s. Indeed, the process $Y_t(\omega, u) := 1\!\!1_{\{R(\omega,u)<t\}}$ is predictable and Proposition 4.18 (a) yields

$$\mathbb{P}(R(\zeta) < t) = \mathbb{E}[Y_t(\zeta)] = \mathbb{E}\left[\int_{\overline{\mathbb{R}}} 1\!\!1_{\{R(u)<t\}}p_{t-}(u)\eta(du)\right] = 0.$$

Remark 4.19 We will often use the notation $\langle p(u), X\rangle_{t\,|u=\zeta}$, $\mathbb{E}[y(u)|\mathscr{F}_t]_{|u=\zeta}$, etc. to denote that we proceed in two steps. In the first step, we compute $\langle p(u), X\rangle_t$ for each $u \in \overline{\mathbb{R}}$. In a second step, we plug $u = \zeta$ in the result. The existence of jointly measurable objects, needed to plug $u = \zeta$, is ensured by the results in Stricker and Yor [205].

Lemma 4.20 *Let ζ be a random variable satisfying Jacod's absolute continuity condition. Then the filtration $\mathbb{F} \vee \sigma(\zeta)$ is right-continuous, i.e., $\mathbb{F}^{\sigma(\zeta)} = \mathbb{F} \vee \sigma(\zeta)$.*

Proof Let Y be a bounded $\mathscr{F}_T \otimes \mathscr{B}(\overline{\mathbb{R}})$-measurable function $(\omega, u) \to Y(\omega, u)$ for some $T > 0$. We shall prove that

$$\mathbb{E}[Y(\zeta)|\mathscr{F}_t \vee \sigma(\zeta)] = \frac{\mathbb{E}[Y(u)p_T(u)|\mathscr{F}_t]_{|u=\zeta}}{p_{t\wedge T}(\zeta)} = y_t(\zeta) \tag{4.11}$$

where $(\omega, t, u) \to y_t(\omega, u) := \frac{\mathbb{E}[Y(u)p_T(u)|\mathscr{F}_t]}{p_{t\wedge T}(u)} 1\!\!1_{\{p_{t\wedge T}(u)>0\}}$ and where we can take a càdlàg and $\mathscr{O}(\mathbb{F}) \otimes \mathscr{B}(\overline{\mathbb{R}})$-measurable version of the $\mathbb{F}$-optional projection of the constant $\mathscr{F}_T \otimes \mathscr{B}(\overline{\mathbb{R}})$-measurable process $(\omega, t, u) \to Y(u)p_T(u)$ (see Proposition 3 in [205] for its existence). Let $t \in [0, T]$, $F \in \mathscr{F}_t$ and h be a bounded Borel function.

Then, by Proposition 4.18 (b) and (4.10), and since $\{p_T(u) > 0\} \subset \{p_t(u) > 0\}$ as $p(u)$ is a non-negative martingale for each $u \in \overline{\mathbb{R}}$, we obtain:

$$\mathbb{E}\left[Y(\zeta)\mathbb{1}_F h(\zeta)\right] = \int_{\overline{\mathbb{R}}} h(u)\mathbb{E}\left[Y(u)\mathbb{1}_F p_T(u)\right]\eta(du)$$

$$= \int_{\overline{\mathbb{R}}} h(u)\mathbb{E}\left[\mathbb{1}_F \mathbb{E}\left[Y(u)p_T(u)|\mathscr{F}_t\right]\mathbb{1}_{\{p_t(u)>0\}}\frac{p_t(u)}{p_t(u)}\right]\eta(du) = \mathbb{E}\left[\mathbb{1}_F h(\zeta)y_t(\zeta)\right]$$

which shows (4.11). By Theorem 1.11, for any $t \in [0, T]$, we have

$$\mathbb{E}\left[Y(\zeta)|\mathscr{F}_t \vee \sigma(\zeta)\right] = y_t(\zeta) = \lim_{s \searrow t} y_s(\zeta) = \lim_{s \searrow t} \mathbb{E}\left[Y(\zeta)|\mathscr{F}_s \vee \sigma(\zeta)\right] = \mathbb{E}\left[Y(\zeta)\middle|\mathscr{F}_t^{\sigma(\zeta)}\right].$$

Since Y and T were chosen arbitrary, we conclude that $\mathscr{F}_t \vee \sigma(\zeta) = \mathscr{F}_t^{\sigma(\zeta)}$ for each $t \geq 0$ which completes the proof. □

Corollary 4.21 *Let ζ be a random variable satisfying Jacod's absolute continuity condition. Then, for $T \geq 0$ and $\mathscr{F}_T \otimes \mathscr{B}(\overline{\mathbb{R}})$-measurable function $(\omega, u) \to Y(\omega, u)$ it holds*

$$\mathbb{E}\left[Y(\zeta)\middle|\mathscr{F}_t^{\sigma(\zeta)}\right] = \frac{\mathbb{E}\left[Y(u)p_T(u)|\mathscr{F}_t\right]_{|u=\zeta}}{p_{t\wedge T}(\zeta)} = y_t(\zeta) \tag{4.12}$$

where $(\omega, t, u) \to y_t(\omega, u)$ is a càdlàg in t and $\mathscr{O}(\mathbb{F}) \otimes \mathscr{B}(\overline{\mathbb{R}})$-measurable function.

Proof Equality (4.12) follows by (4.11) and right-continuity of $\mathbb{F} \vee \sigma(\zeta)$. □

Proposition 4.22 *Let ζ be a r.v. satisfying Jacod's absolute continuity condition.*
(a) A random variable Y is $\mathscr{F}_t^{\sigma(\zeta)}$-measurable if and only if Y is of the form $Y(\omega) = y(\omega, \zeta(\omega))$ for some $\mathscr{F}_t \otimes \mathscr{B}(\overline{\mathbb{R}})$-measurable function $(\omega, u) \to y(\omega, u)$.
(b) A process Y is $\mathbb{F}^{\sigma(\zeta)}$-optional if and only if $Y_t(\omega)$ is a modification of $y_t(\omega, \zeta(\omega))$ where $(\omega, t, u) \to y_t(\omega, u)$ is an $\mathscr{O}(\mathbb{F}) \otimes \mathscr{B}(\overline{\mathbb{R}})$-measurable function.
(c) A process Y is $\mathbb{F}^{\sigma(\zeta)}$-predictable if and only if $Y_t(\omega)$ is a modification of $y_t(\omega, \zeta(\omega))$ where $(\omega, t, u) \mapsto y_t(\omega, u)$ is a $\mathscr{P}(\mathbb{F}) \otimes \mathscr{B}(\overline{\mathbb{R}})$-measurable function.
(d) Let y be an $\mathscr{F} \otimes \mathscr{B}(\overline{\mathbb{R}})$-measurable function. Then, for any t,

$$\mathbb{E}\left[y(\zeta)\middle|\mathscr{F}_t^{\sigma(\zeta)}\right] = \mathbb{E}\left[y(u)\middle|\mathscr{F}_t^{\sigma(\zeta)}\right]_{|u=\zeta}.$$

Proof (a) Sufficiency of the condition follows from measurability of function composition. To prove necessity we use Lemma 4.20. Let Y be an $\mathscr{F}_t \vee \sigma(\zeta)$-measurable r.v. By the monotone class theorem, we will show that Y is necessarily of the form $y_t(\omega, \zeta(\omega))$ for some $\mathscr{F}_t \otimes \mathscr{B}(\overline{\mathbb{R}})$-measurable function $(\omega, u) \to y_t(\omega, u)$. Let $\mathscr{C} = \{F \cap G : F \in \mathscr{F}_t \text{ and } G \in \sigma(\zeta)\}$ and note that $\mathscr{C}$ is stable under finite intersection. Define a class of random variables $\mathscr{H}$ by

$$\mathscr{H} = \{Y : \exists \mathscr{F}_t \otimes \mathscr{B}(\overline{\mathbb{R}})\text{-measurable } y \text{ such that } Y(\omega) = y(\omega, \zeta(\omega))\}.$$

Note that $\mathscr{H}$ satisfies: $1 \in \mathscr{H}$, $\mathbb{1}_F \in \mathscr{H}$ for any $F \in \mathscr{C}$ and for any sequence $(Y_n)_{n\geq 1} \in \mathscr{H}$ such that Y_n increases to a r.v Y and $Y < \infty$ then $Y \in \mathscr{H}$ since $Y_n(\omega) = y_n(\omega, \zeta(\omega)) = \widetilde{y}_n(\omega, \zeta(\omega))$ where $\widetilde{y}_n(\omega, u) = \mathbb{1}_{\{u=\zeta(\omega)\}} y_n(\omega, u)$ is also $\mathscr{F}_t \otimes \mathscr{B}(\overline{\mathbb{R}})$-measurable and the sequence $(\widetilde{y}_n(\omega, u))_{n\geq 1}$ converges to a finite limit $y(\omega, u)$ which is $\mathscr{F}_t \otimes \mathscr{B}(\overline{\mathbb{R}})$-measurable thus $Y(\omega) = y(\omega, \zeta(\omega))$. Therefore $\mathscr{H}$ contains all $\sigma(\mathscr{C})$-measurable random variables.

(b) Similarly as in Proposition 2.11 it is enough to establish the decomposition for any u.i. càdlàg $\mathbb{F}^{\sigma(\zeta)}$-martingale Y. Since $\mathbb{E}\left[Y_n \middle| \mathscr{F}_t^{\sigma(\zeta)}\right] = Y_t$ holds for all $t \leq n$, for every $n \in \mathbb{N}$, by Corollary 4.21 and assertion (a), there exists an $\mathscr{O}(\mathbb{F}) \otimes \mathscr{B}(\overline{\mathbb{R}})$-measurable function $(\omega, t, u) \to y_t^n(\omega, u)$ such that $Y_t = y_t^n(\zeta)$ for all $t \in [0, n]$ and for every $n \in \mathbb{N}$. Moreover, it holds that $y_t^{n+1}(\zeta) = y_t^n(\zeta)$ for $t \leq n$ which completes the proof since it is enough to consider $y_t(u) = \sum_{n=1}^{\infty} \mathbb{1}_{[n-1,n)} y_t^n(u)$.

(c) Let Y be càglàd and $\mathbb{F}^{\sigma(\zeta)}$-adapted. Therefore, by assertion (a), for any fixed t there exists an $\mathscr{F}_t \otimes \mathscr{B}(\overline{\mathbb{R}})$-measurable function $(\omega, u) \to y_t(\omega, u)$ such that $Y_t(\omega) = y_t(\omega, \zeta(\omega))$. To guarantee enough regularity, let $\widetilde{y}_t(\omega, u) = \mathbb{1}_{\{u=\zeta(\omega)\}} y_t(\omega, u)$ so that $y_t(\omega, \zeta(\omega)) = \widetilde{y}_t(\omega, \zeta(\omega))$ and note that for $\mathbb{P}$-a.e. ω and each u the path $(\widetilde{y}(\omega, u))_{t\geq 0}$ is càglàd since $\widetilde{y}(\omega, u) = 0$ if $u \neq \zeta(\omega)$ and $\widetilde{y}(\omega, u) = Y(\omega)$ if $u = \zeta(\omega)$. Hence $\widetilde{y}$ is $\mathscr{P}(\mathbb{F} \otimes \mathscr{B}(\overline{\mathbb{R}}))$-measurable. The conclusion follows from $\mathscr{P}(\mathbb{F} \otimes \mathscr{B}(\overline{\mathbb{R}})) = \mathscr{P}(\mathbb{F}) \otimes \mathscr{B}(\overline{\mathbb{R}})$.

(d) Let y be of the form $y(\omega, u) = g(u)Y(\omega)$ where Y is an $\mathscr{F}$-measurable r.v. and g is a Borel function. Then, $y(\zeta) = g(\zeta)Y$ and we have

$$\mathbb{E}\left[y(\zeta) \middle| \mathscr{F}_t^{\sigma(\zeta)}\right] = g(\zeta)\mathbb{E}\left[Y \middle| \mathscr{F}_t^{\sigma(\xi)}\right] = \left(\mathbb{E}\left[y(u) \middle| \mathscr{F}_t^{\sigma(\xi)}\right]\right)_{|u=\zeta}.$$

For a general y, we proceed by the monotone class theorem. □

Lemma 4.23 *Assume that ζ satisfies Jacod's equivalence condition. Then, for an $\mathscr{O}(\mathbb{F}) \otimes \mathscr{B}(\overline{\mathbb{R}})$-measurable function $(\omega, t, u) \to Y_t(\omega, u)$, one has $Y(\zeta) = 0$ $\mathbb{P}$-a.s. if and only if, for η-a.e. u, $Y(u) = 0$ $\mathbb{P}$-a.s.*

Proof The sufficient condition is straightforward. The necessary condition follows by Proposition 4.18 (b) combined with positivity of $p(u)$ for η-a.e. u

$$0 = {}^{o,\mathbb{F}}(|Y(\zeta)|)_t = \int_{\overline{\mathbb{R}}} |Y_t(u)| p_t(u)\eta(du), \quad t \geq 0.$$

□

Remark 4.24 Let us write the result of Proposition 4.9 in terms of Jacod's absolute continuity condition. If Jacod's absolute continuity condition is satisfied, then

$$\lambda_t(f) = \int_{\overline{\mathbb{R}}} p_t(u) f(u)\eta(du).$$

Hence, on the one hand $\langle \lambda(f), X\rangle_t = \int_0^t \widehat{\lambda}_s(f)ds = \int_{\overline{\mathbb{R}}} f(u)\, \langle p(u), X\rangle_t \eta(du)$ and on the other hand $\int_0^t \int_{\overline{\mathbb{R}}} f(u)\widehat{\lambda}_s(du)ds = \int_0^t \int_{\overline{\mathbb{R}}} f(u)\, \langle p(u), X\rangle_s \eta(du)ds$. Therefore, $\rho(\zeta, t)dt = \left(\frac{d\langle p(u), X\rangle_t}{p_t(u)}\right)_{|u=\zeta}$. We will see in Theorem 4.25 that this quantity is well-defined.

4.4.2 Hypothesis $\mathscr{H}'$ and Semimartingale Decomposition

In the next theorem the hypothesis $\mathscr{H}'$ under Jacod's absolute continuity condition is proven, and the $\mathbb{F}^{\sigma(\zeta)}$-canonical semimartingale decomposition of $\mathbb{F}$-local martingales is provided. Then, in Theorem 4.28, another $\mathbb{F}^{\sigma(\zeta)}$-semimartingale decomposition of $\mathbb{F}$-local martingales is shown.

Theorem 4.25 *Suppose that ζ satisfies Jacod's absolute continuity condition. If X is an $\mathbb{F}$-local martingale, the process $\widetilde{X}$ defined as*

$$\widetilde{X}_t := X_t - \int_0^t \frac{1}{p_{s-}(\zeta)} d\langle X, p(u)\rangle_s^{\mathbb{F}}{}_{|u=\zeta}\,, \quad t \leq T$$

is an $\mathbb{F}^{\sigma(\zeta)}$-local martingale. In particular, the hypothesis $\mathscr{H}'$ is satisfied.

Proof By Proposition 4.17 (a) the processes $(p(u), u \in \overline{\mathbb{R}})$ are $\mathbb{F}$-martingales. We give the proof in the case when the $\mathbb{F}$-martingales X and $(p(u), u \in \overline{\mathbb{R}})$ are locally square integrable, for the general case, we refer the reader to Jacod [123, Theorem 2.5]. Since X and $p(u)$ are square integrable martingales, by Example 1.38, $\langle X, p(u)\rangle^{\mathbb{F}}$ exists. Let F_s be a bounded $\mathscr{F}_s$-measurable random variable and $h : \overline{\mathbb{R}} \to \mathbb{R}$ be a bounded Borel function. Then the variable $F_s h(\zeta)$ is $\mathscr{F}_s^{\sigma(\zeta)}$-measurable and if a decomposition of the form $X_t = \widetilde{X}_t + K_t(\zeta)$ holds, the $\mathbb{F}^{\sigma(\zeta)}$-martingale property of $\widetilde{X}$ implies, for $t \geq s$, $\mathbb{E}\left[F_s h(\zeta)(\widetilde{X}_t - \widetilde{X}_s)\right] = 0$, hence

$$\mathbb{E}\left[F_s h(\zeta)\,(X_t - X_s)\right] = \mathbb{E}\left[F_s h(\zeta)(K_t(\zeta) - K_s(\zeta))\right]. \tag{4.13}$$

We can write:

$$\begin{aligned}
\mathbb{E}\left[F_s h(\zeta)\,(X_t - X_s)\right] &= \mathbb{E}\left[F_s\,(X_t - X_s) \int_{\overline{\mathbb{R}}} h(u)\, p_t(u)\eta(du)\right] \\
&= \int_{\overline{\mathbb{R}}} h(u)\mathbb{E}\left[F_s\,(X_t p_t(u) - X_s p_s(u))\right]\eta(du) \\
&= \int_{\overline{\mathbb{R}}} h(u)\mathbb{E}\left[F_s \int_s^t d\,\langle X, p(u)\rangle_v^{\mathbb{F}}\right]\eta(du) \qquad (4.14)
\end{aligned}$$

where the first equality comes from conditioning on $\mathscr{F}_t$, the second from the martingale property of $p(u)$ given in Proposition 4.17 (a), and the third from integration by parts and martingale property of $[X, p(u)] - \langle X, p(u)\rangle^{\mathbb{F}}$. For a fixed u define the $\mathbb{F}$-stopping time $\widetilde{R}(u) := R(u)_{\{p_{R(u)-}(u)>0\}}$ where $R(u)$ was introduced in (4.7). Note that, by Proposition 1.36 (b) and Example 1.33,

$$\Delta\langle X, p(u)\rangle^{\mathbb{F}}_{\widetilde{R}(u)} = {}^{p,\mathbb{F}}\left(\Delta[X, p(u)]_{\widetilde{R}(u)}\right) = -p_{\widetilde{R}(u)-}(u)\,{}^{p,\mathbb{F}}\left(\Delta X_{\widetilde{R}(u)}\right) = 0.$$

Then, since, by Proposition 4.17, $p(u) = 0$ and $p_-(u) = 0$ on $]\!]R(u), \infty[\![$, one has $\mathbb{1}_{\{p_-(u)=0\}} \cdot \langle X, p(u)\rangle^{\mathbb{F}} = \mathbb{1}_{[\![\widetilde{R}(u)]\!]} \cdot \langle X, p(u)\rangle^{\mathbb{F}} = 0$. Applying the latter fact to the computations of (4.14), we obtain

$$\begin{aligned}
(4.14) &= \int_{\overline{\mathbb{R}}} h(u)\mathbb{E}\left[F_s \int_s^t \frac{p_{v-}(u)}{p_{v-}(u)} \mathbb{1}_{\{p_{v-}(u)>0\}}\, d\,\langle X, p(u)\rangle^{\mathbb{F}}_v\right]\eta(du)\\
&= \int_{\overline{\mathbb{R}}} h(u)\mathbb{E}\left[F_s p_t(u) \int_s^t \frac{1}{p_{v-}(u)} \mathbb{1}_{\{p_{v-}(u)>0\}}\, d\,\langle X, p(u)\rangle^{\mathbb{F}}_v\right]\eta(du)\\
&= \mathbb{E}\left[F_s \int_{\overline{\mathbb{R}}} h(u)p_t(u) \int_s^t \frac{1}{p_{v-}(u)} \mathbb{1}_{\{p_{v-}(u)>0\}}\, d\,\langle X, p(u)\rangle^{\mathbb{F}}_v\,\eta(du)\right]\\
&= \mathbb{E}\left[F_s h(\zeta) \int_s^t \frac{1}{p_{v-}(\zeta)} d\langle X, p(u)\rangle^{\mathbb{F}}_{v}\,{}_{|u=\zeta}\right]
\end{aligned}$$

where the second equality comes from the martingale property of $p(u)$ and integration by parts, and the last equality from Proposition 4.18 (a) and (4.7). Comparing this to the right-hand side in (4.13), we obtain that $dK(u) = d\,\langle X, p(u)\rangle^{\mathbb{F}} / p_-(u)$. □

Remark 4.26 In the proof of the above theorem we only considered the case where X and $p(u)$ are locally square integrable martingales. That allowed us to establish the existence of $\langle X, p(u)\rangle^{\mathbb{F}}$. In a general case it is much more difficult. We refer to [123, Theorem 2.5 (a)] where the existence of the predictable covariation process $\langle X, p(u)\rangle^{\mathbb{F}}$ for η-a.e. u, is proven. More precisely [123, Theorem 2.5 (a)] establishes that under Jacod's absolute continuity condition, for every $\mathbb{F}$-(local)martingale X, there exists a η-negligible set B (depending on X), such that $\langle X, p(u)\rangle^{\mathbb{F}}$ is well-defined for $u \notin B$. Having that result, similar computations can be done in a general case.

Before showing an optional decomposition in Theorem 4.28, recall the following result (proven in [10]):

Lemma 4.27 *Suppose that ζ satisfies Jacod's absolute continuity condition. Let $V : (\omega, t, u) \to \mathbb{R}$ be $\mathscr{O}(\mathbb{F}) \otimes \mathscr{B}(\overline{\mathbb{R}})$-measurable such that for each u, $V(u)$ is a càdlàg process with locally integrable variation. Then the process $U(u) := \frac{1}{p(\zeta)} \cdot V(u)$ is well-defined, its variation is $\mathbb{F}^{\sigma(\zeta)}$-locally integrable, and the $\mathbb{F}^{\sigma(\zeta)}$-dual predictable projection of $U(\zeta)$ is given by*

$$(U(\zeta))^{p,\mathbb{F}^{\sigma(\zeta)}} = \frac{1}{p_-(\zeta)} \cdot \left(\mathbb{1}_{\{p(u)>0\}} \cdot V(u)\right)^{p,\mathbb{F}}{}_{|u=\zeta}.$$

Theorem 4.28 *Suppose that ζ satisfies Jacod's absolute continuity condition. Let X be an $\mathbb{F}$-local martingale. Then,*

$$\bar{X}_t := X_t - \int_0^t \frac{1}{p_s(\zeta)} d[X, p(\zeta)]_s + \left(\mathbb{1}_{[\![\widetilde{R}(u),\infty[\![} \Delta X_{\widetilde{R}(u)}\right)_t^{p,\mathbb{F}}\ _{|u=\zeta} \tag{4.15}$$

is an $\mathbb{F}^{\sigma(\zeta)}$-local martingale. Here, $\widetilde{R}(u) := R(u)_{\{p_{R(u)-}(u)>0\}}$ with $R(u)$ defined in (4.7).

Proof From the predictable decomposition given in Theorem 4.25 and Lemma 4.27, we expand X as

$$\begin{aligned} X &= \widetilde{X} + \frac{1}{p_-(\zeta)}\left(\mathbb{1}_{\{p(u)>0\}} \cdot [X, p(u)]\right)^{p,\mathbb{F}}_{|u=\zeta} + \frac{1}{p_-(\zeta)} \cdot \left(\mathbb{1}_{\{p(u)=0\}} \cdot [X, p(u)]\right)^{p,\mathbb{F}}_{|u=\zeta} \\ &= \widetilde{X} + \left(\frac{1}{p(\zeta)} \cdot [X, p(\zeta)]\right)^{p,\mathbb{F}^{\sigma(\zeta)}} + \frac{1}{p_-(\zeta)} \cdot \left(\Delta X_{\widetilde{R}(u)}\, \Delta p_{\widetilde{R}(u)}(u) \mathbb{1}_{[\![\widetilde{R}(u),\infty[\![}\right)^{p,\mathbb{F}}_{|u=\zeta} \\ &= \bar{X} + \frac{1}{p(\zeta)} \cdot [X, p(\zeta)] - \left(\Delta X_{\widetilde{R}(u)} \mathbb{1}_{[\![\widetilde{R}(u),\infty[\![}\right)^{p,\mathbb{F}}_{|u=\zeta} \end{aligned}$$

where

$$\bar{X} := \widetilde{X} - \frac{1}{p(\zeta)} \cdot [X, p(\zeta)] + \left(\frac{1}{p(\zeta)} \cdot [X, p(\zeta)]\right)^{p,\mathbb{F}^{\sigma(\zeta)}}$$

is an $\mathbb{F}^{\sigma(\zeta)}$-local martingale. □

An important application of the decomposition established in Theorem 4.28 concerns the predictable representation property (PRP). More precisely, if X has the PRP in $\mathbb{F}$, then, $\bar{X}$ has the PRP in $\mathbb{F}^{\sigma(\zeta)}$. This is presented in detail in Sect. 4.4.3.

Corollary 4.29 *(a) The process $\frac{1}{p(\zeta)}$ is an $\mathbb{F}^{\sigma(\zeta)}$-supermartingale with decomposition*

$$\frac{1}{p(\zeta)} = \frac{1}{p_0(\zeta)} - \frac{1}{(p_-(\zeta))^2} \cdot \bar{p}(\zeta) - \frac{1}{p_-(\zeta)} \cdot \left(\mathbb{1}_{[\![\widetilde{R}(u),\infty[\![}\right)^{p,\mathbb{F}}_{|u=\zeta}, \tag{4.16}$$

where $\bar{p}(\zeta)$ is the $\mathbb{F}^{\sigma(\zeta)}$-local martingale given by (4.15)*. Moreover, it can be written as a stochastic exponential of the form*

$$\frac{p_0(\zeta)}{p(\zeta)} = \mathscr{E}\left(-\frac{1}{p_-(\zeta)} \cdot \bar{p}(\zeta) - \left(\mathbb{1}_{[\![\widetilde{R}(u),\infty[\![}\right)^{p,\mathbb{F}}_{|u=\zeta}\right).$$

(b) The process $\frac{1}{p(\zeta)}$ is an $\mathbb{F}^{\sigma(\zeta)}$-local martingale if and only if $\widetilde{R}(u) = \infty$ $\mathbb{P} \otimes \eta$-a.s. Then $\frac{p_0(\zeta)}{p(\zeta)} = \mathscr{E}(-\frac{1}{p_-(\zeta)} \cdot \bar{p}(\zeta))$.

Proof (a) Since $p : (\omega, t, u) \to \mathbb{R}$ is $\mathcal{O}(\mathbb{F}) \otimes \mathcal{B}(\overline{\mathbb{R}})$-measurable and $p(u)$ is an $\mathbb{F}^{\sigma(\zeta)}$-semimartingale for each u, the process $p(\zeta)$ is an $\mathbb{F}^{\sigma(\zeta)}$-semimartingale. Itô's lemma implies that

$$\frac{1}{p(\zeta)} = \frac{1}{p_0(\zeta)} - \frac{1}{(p_-(\zeta))^2} \cdot p(\zeta) + \frac{1}{(p_-(\zeta))^2 p(\zeta)} \cdot [p(\zeta)].$$

Then the decomposition (4.16) follows by Theorem 4.28 combined with Proposition 4.22 (d) and the supermartingale property is obtained. The exponential form is immediate.

(b) The process $\frac{1}{p(\zeta)}$ is an $\mathbb{F}^{\sigma(\zeta)}$-local martingale if and only if $\left(\mathbb{1}_{[\![\widetilde{R}(u),\infty[\![}\right)^{p,\mathbb{F}}|_{u=\zeta}$ is null. The last is equivalent to knowing that, for each t

$$\begin{aligned} 0 &= \mathbb{E}\left[(\mathbb{1}_{[\![\widetilde{R}(u),\infty[\![})^{p,\mathbb{F}}_t |_{u=\zeta}\right] = \mathbb{E}\left[\,{}^{o,\mathbb{F}}\left((\mathbb{1}_{[\![\widetilde{R}(u),\infty[\![})^{p,\mathbb{F}} |_{u=\zeta}\right)_t\,\right] \\ &= \mathbb{E}\left[\int_{\overline{\mathbb{R}}} (\mathbb{1}_{[\![\widetilde{R}(u),\infty[\![})^{p,\mathbb{F}}_t p_t(u)\eta(du)\right] = \int_{\overline{\mathbb{R}}} \mathbb{E}\left[(\mathbb{1}_{[\![\widetilde{R}(u),\infty[\![})^{p,\mathbb{F}}_t p_t(u)\right]\eta(du), \end{aligned}$$

where the second equality comes from (4.9). By Yoeurp's lemma we conclude that, for each t

$$0 = \int_{\overline{\mathbb{R}}} \mathbb{E}\left[\int_0^t p_{s-}(u) d\left((\mathbb{1}_{[\![\widetilde{R}(u),\infty[\![})^{p,\mathbb{F}}\right)_s\right]\eta(du) = \int_{\overline{\mathbb{R}}} \mathbb{E}\left[p_{\widetilde{R}(u)-}(u)\mathbb{1}_{\{\widetilde{R}(u)\le t\}}\right]\eta(du)$$

which in turn is equivalent to $\widetilde{R}(u) > t$ $\mathbb{P} \otimes \eta$-a.s. for each t since $p_{\widetilde{R}(u)-}(u) > 0$. Thus, $\frac{1}{p(\zeta)}$ is an $\mathbb{F}^{\sigma(\zeta)}$-local martingale if and only if $\widetilde{R}(u)$ is infinite $\mathbb{P} \otimes \eta$-a.s. □

Example 4.30 (*Atomic σ-fields*) We come back to Example 4.15, i.e., we consider an initial enlargement with a random variable ζ which takes countably many values $(c_n)_n$ or, equivalently, an enlargement with an atomic σ-field. For $C_n = \{\zeta = c_n\}$ where $\mathbb{P}(C_n) > 0$, the sequence $(C_n)_n$ forms a partition of Ω. Then, for each n, we define z^n as the $\mathbb{F}$-martingale $z^n_t := \mathbb{P}(C_n|\mathcal{F}_t)$. We have $\{\inf_s z^n_s = 0\} \cap C_n = \emptyset$. Let us now denote by $\mathbb{F}^{\mathcal{C}}$ the initial enlargement of the filtration $\mathbb{F}$ with the atomic σ-field $\mathcal{C} = \sigma((C_n)_n) = \sigma(\zeta)$. In this case, as already shown in Example 4.15, Jacod's absolute continuity condition is satisfied with $p_t(u) = \sum_{n=1}^{\infty} \frac{z^n_t}{z^n_0} \mathbb{1}_{\{u=c_n\}}$, thus the hypothesis $\mathcal{H}'$ holds and each $\mathbb{F}$-martingale X can be decomposed in $\mathbb{F}^{\mathcal{C}}$ as

$$X_t = \widehat{X}_t + \sum_n \mathbb{1}_{C_n} \int_0^t \frac{1}{z^n_{s-}} d\langle X, z^n\rangle^{\mathbb{F}}_s,$$

where $\widehat{X}$ is an $\mathbb{F}^{\mathcal{C}}$-local martingale.

A special case of the previous example is the Poisson bridge. Indeed, for N a Poisson process and $\zeta = N_T$, one has

$$p_t(u) = \sum_{n=0}^{\infty} e^t \frac{(T-t)^{n-N_t}}{T^n} \frac{n!}{(n-N_t)!} \mathbb{1}_{\{N_t \le n\}} \mathbb{1}_{\{u=n\}}.$$

Example 4.31 (*Enlargement with* $\int_0^\infty f(s)dB_s$) As shown in Example 4.16, Jacod's equivalence condition is satisfied in the case where $\zeta = \int_0^\infty f(s)dB_s$ for a Brownian motion B and its natural filtration $\mathbb{F}$. Note that, from Itô's calculus,

$$dp_t(u) = p_t(u)\frac{u - x_t}{\sigma^2(t)}dx_t\,,$$

hence, $d\langle p(u), B\rangle_{t|u=\zeta} = p_t(\zeta)\frac{1}{\sigma^2(t)}(\zeta - x_t)f(t)dt$.

We obtain that B is an $\mathbb{F}^{\sigma(\zeta)}$-semimartingale with canonical decomposition:

$$B_t = \widetilde{B}_t + \int_0^t \frac{f(s)}{\sigma^2(s)}\left(\int_s^\infty f(u)\,dB_u\right)ds.$$

4.4.3 Predictable Representation Property

We denote by $\mathcal{M}^2(\mathbb{F})$ the set of $\mathbb{F}$-martingales x, such that

$$\mathbb{E}\left[x_t^2\right] < \infty, \quad \forall\, t \ge 0. \tag{4.17}$$

For an $\mathbb{F}$-local martingale x, we denote by $\mathcal{L}(x,\mathbb{F})$ the set of $\mathbb{F}$-predictable processes which are integrable with respect to x (in the sense of local martingale), namely (see, e.g., Definition 9.1 and Theorem 9.2. in [112])

$$\mathcal{L}(x,\mathbb{F}) = \left\{\varphi \in \mathcal{P}(\mathbb{F}) : \left(\int_0^\cdot \varphi_s^2 d[x]_s\right)^{1/2} \text{ is locally integrable}\right\}.$$

Hypothesis 4.32 There exists a process $x \in \mathcal{M}_{\text{loc}}(\mathbb{F})$ such that every $y \in \mathcal{M}_{\text{loc}}(\mathbb{F})$ can be represented as $y = y_0 + \varphi \bullet x$ for some $\varphi \in \mathcal{L}(x,\mathbb{F})$.

In the next proposition we characterize $\mathbb{F}^{\sigma(\zeta)}$-martingales in terms of $\mathbb{F}$-martingales.

Proposition 4.33 *Assume that* ζ *satisfies Jacod's absolute continuity condition. Then, a process* Y *is an* $\mathbb{F}^{\sigma(\zeta)}$*-martingale if and only if there exists an* $\mathcal{O}(\mathbb{F}) \otimes \mathcal{B}(\overline{\mathbb{R}})$*-measurable function* $(\omega, t, u) \to y_t(\omega, u)$ *such that* $Y = y(\zeta)$ *and* $y(u)p(u)$ *is an* $\mathbb{F}$*-martingale for every* $u \in \overline{\mathbb{R}}$.

Proof The proof relies on similar arguments to the proof of Proposition 4.17. To show the necessary condition let Y be an $\mathbb{F}^{\sigma(\zeta)}$-martingale. Then, by Proposition 4.22 (b), there exists an $\mathcal{O}(\mathbb{F}) \otimes \mathcal{B}(\overline{\mathbb{R}})$-measurable function $(\omega, t, u) \to \widehat{y}_t(\omega, u)$, such that $Y_t = y_t(\zeta)$ a.s. for all $t \geq 0$ and, for $s \leq t$ with $s, t \in \mathbb{Q}^+$, a bounded $\mathscr{F}_s$-measurable r.v. X and a bounded Borel function g, we have

$$\begin{aligned}\mathbb{E}[Y_t X g(\zeta)] = \mathbb{E}[\widehat{y}_t(\zeta) X g(\zeta)] &= \int_{\overline{\mathbb{R}}} g(u)\mathbb{E}[\widehat{y}_t(u) p_t(u) X]\,\eta(du) \\ &= \int_{\overline{\mathbb{R}}} g(u)\mathbb{E}[X\mathbb{E}[\widehat{y}_t(u) p_t(u)|\mathscr{F}_s]]\,\eta(du) \qquad (4.18)\end{aligned}$$

where, similarly as in the proof of Proposition 4.17, we take an $\mathscr{F}_s \otimes \mathscr{B}(\overline{\mathbb{R}})$-measurable version of $\mathbb{E}[\widehat{y}_t(u) p_t(u)|\mathscr{F}_s]$ (see Lemma 3 in [205] for existence). From the martingale property of Y and the equality (4.18) we obtain that the set $\{(\omega, u) : \widehat{y}_s(u) p_s(u) \neq \mathbb{E}[\widehat{y}_t(u) p_t(u)|\mathscr{F}_s]\}$ is $\mathbb{P} \otimes \eta$-negligible. Continuing as in the proof of Proposition 4.17, we conclude the existence of an $\mathscr{O}(\mathbb{F}) \otimes \mathscr{B}(\overline{\mathbb{R}})$-measurable function $(\omega, t, u) \to y_t(\omega, u)$ such that $Y_t(\zeta) = y_t(\zeta)$ a.s. for any $t \geq 0$ and $y(u)p(u)$ is an $\mathbb{F}$-martingale for all $u \in \overline{\mathbb{R}}$.

The sufficient condition follows again by (4.18). □

Theorem 4.34 *Assume that ζ satisfies Jacod's absolute continuity condition. Under Hypothesis 4.32, every $Y \in \mathscr{M}_{\text{loc}}(\mathbb{F}^{\sigma(\zeta)})$ admits a representation*

$$Y_t = Y_0 + \int_0^t \Phi_s d\bar{x}_s \qquad (4.19)$$

where $\Phi \in \mathscr{L}(\bar{x}, \mathbb{F}^{\sigma(\zeta)})$ and $Y_0 \in \mathscr{F}_0^{\sigma(\zeta)}$. Here $\bar{x}$ is the $\mathbb{F}^{\sigma(\zeta)}$-martingale part (given in (4.15)) of the $\mathbb{F}^{\sigma(\zeta)}$-semimartingale x introduced in Hypothesis 4.32.

Sketch of the proof: (For the full proof, we refer to Theorem 2.6 in [99].) Let Y be a u.i. $\mathbb{F}^{\sigma(\zeta)}$-martingale. By Proposition 4.33 one has $Y = y(\zeta)$ for an $\mathscr{O}(\mathbb{F}) \otimes \mathscr{B}(\overline{\mathbb{R}})$-measurable function $(\omega, t, u) \to y_t(\omega, u)$ such that $Y = y(\zeta)$ and $y(u)p(u)$ is an $\mathbb{F}$-martingale, for η-almost every u. Thus $Y_t = \frac{y_t(\zeta)p_t(\zeta)}{p_t(\zeta)}$. Then, as $y(u)p(u)$ and $p(u)$ are $\mathbb{F}$-martingales for each u, they have a representation w.r.t. x. This representation can be chosen to be appropriately measurable (which is a delicate issue), namely $y(u)p(u) = y_0(u)p_0(u) + \varphi(u) \cdot x$ and $p(u) = p_0(u) + \psi(u) \cdot x$ where φ and ψ are $\mathscr{P}(\mathbb{F}) \otimes \mathscr{B}(\overline{\mathbb{R}})$-measurable. Then, the result follows by integration by parts using the decomposition (4.15) of an $\mathbb{F}$-local martingale x. □

4.4.4 NUPBR and Deflators

In this subsection, we focus on the NUPBR in an initial enlargement framework under Jacod's absolute continuity condition. We give, without proofs, some general results (see [10] for details and more information). In Proposition 4.35, we give $\mathbb{F}^{\sigma(\zeta)}$-local martingale deflators for quasi left-continuous $\mathbb{F}$-local martingales and $\mathbb{F}^{\sigma(\zeta)}$-supermartingale deflators for $\mathbb{F}$-local martingales. We recall that a process X is quasi-left-continuous if it does not jump at predictable stopping times. Proposition 4.35

can be shown by Theorem 4.28 and integration by parts. In Theorem 4.36, we present a sufficient condition such that any $\mathbb{F}$-local martingale satisfies NUPBR($\mathbb{F}^{\sigma(\zeta)}$).

Recall that $\widetilde{R}(u) := R(u)_{\{p_{R(u)-}(u)>0\}}$ where $R(u) := \inf\{t : p_t(u) = 0\}$.

Proposition 4.35 *Let X be an $\mathbb{F}$-local martingale satisfying $\Delta X_{\widetilde{R}(u)} = 0$ on the set $\{\widetilde{R}(u) < \infty\}$ $\mathbb{P} \otimes \eta$-a.s.*
(a) If X is $\mathbb{F}$-quasi-left continuous, then the process $Y := \mathscr{E}\left(-\frac{1}{p_-(\zeta)} \cdot \bar{p}(\zeta)\right)$ is an $\mathbb{F}^{\sigma(\zeta)}$-local martingale deflator for X.
(b) In general, the process $\widetilde{Y} := \mathscr{E}\left(-\frac{1}{p_-(\zeta)} \cdot \bar{p}(\zeta) - V\right)$ is an $\mathbb{F}^{\sigma(\zeta)}$-supermartingale deflator for X, with V defined by

$$V_t := \sum_{0\le s\le t} {}^{p,\mathbb{F}}\left(\mathbb{1}_{[\![\widetilde{R}(u)]\!]}\right)_s |_{u=\zeta}.$$

Theorem 4.36 *If the following equivalent conditions hold:*
(a) the thin set $\{p(u) = 0 < p_-(u)\}$ is evanescent η-a.e.,
(b) the $\mathbb{F}$-stopping time $\widetilde{R}(u)$ is infinite $\mathbb{P} \otimes \eta$-a.s.
Then, any $\mathbb{F}$-local martingale X admits the $\mathbb{F}^{\sigma(\zeta)}$-local martingale deflator $\frac{1}{p(\zeta)}$ and satisfies NUPBR($\mathbb{F}^{\sigma(\zeta)}$).

Proof This is a consequence of Proposition 4.35 and Corollary 4.29. □

4.5 Jacod's Equivalence Condition

Although Jacod's equivalence condition is just a special case of the Jacod's absolute continuity condition, we now concentrate on it particularly. The reason is that, in this situation, there exists a probability measure equivalent to $\mathbb{P}$ under which ζ and $\mathbb{F}$ are independent. This feature significantly simplifies our proofs. Returning to the original probability measure is achieved by means of Girsanov's theorem. Jacod's equivalence condition is also called the hypothesis $\mathscr{E}$ in the literature.

Under the hypothesis $\mathscr{E}$, the family of densities $(p(u), u \in \overline{\mathbb{R}})$ from Proposition 4.17 consists of positive elements.

4.5.1 An Important Change of Probability Measure

Theorem 4.37 *Assume that ζ satisfies hypothesis $\mathscr{E}$. Then the process $L := \frac{p_0(\zeta)}{p(\zeta)}$ is a positive $(\mathbb{F}^{\sigma(\zeta)}, \mathbb{P})$-martingale, with $L_0 = 1$. There exists a probability measure $\mathbb{P}^*$ on $\mathscr{F}_\infty^{\sigma(\zeta)}$ such that, for any $t \ge 0$, $\mathbb{P}^*$ is equivalent to $\mathbb{P}$ on $\mathscr{F}_t^{\sigma(\zeta)}$ and satisfies*

$$d\mathbb{P}^*|_{\mathscr{F}_t^{\sigma(\zeta)}} = L_t\, d\mathbb{P}|_{\mathscr{F}_t^{\sigma(\zeta)}};\quad \mathbb{P}^*|_{\mathscr{F}_t} = \mathbb{P}|_{\mathscr{F}_t} \quad \text{and} \quad \mathbb{P}^*|_{\sigma(\zeta)} = \mathbb{P}|_{\sigma(\zeta)}.$$

Moreover, under $\mathbb{P}^*$, *the random variable* ζ *and the* σ*-field* $\mathscr{F}_\infty$ *are conditionally independent given* $\mathscr{F}_0$.

Proof Obviously, one has $L_0 := \frac{p_0(\zeta)}{p_0(\zeta)} = 1$. Setting $L_t(u) := \frac{p_0(u)}{p_t(u)}$, Proposition 4.18 (b) implies that, for any bounded Borel function $h : \overline{\mathbb{R}} \to \mathbb{R}$, any $\mathscr{F}_s$-measurable bounded random variable K_s and $s \le t$, one has

$$\begin{aligned}\mathbb{E}[L_t h(\zeta) K_s | \mathscr{F}_0] &= \mathbb{E}\left[K_s \int_{\overline{\mathbb{R}}} L_t(u) h(u) p_t(u) \eta(du) \Big| \mathscr{F}_0\right] \\ &= \mathbb{E}\left[K_s \int_{\overline{\mathbb{R}}} h(u) p_0(u) \eta(du) \Big| \mathscr{F}_0\right] \\ &= \mathbb{E}[K_s|\mathscr{F}_0] \int_{\overline{\mathbb{R}}} h(u) p_0(u) \eta(du) = \mathbb{E}[K_s|\mathscr{F}_0]\, \mathbb{E}[h(\zeta)|\mathscr{F}_0].\end{aligned}$$

Hence, the arbitrariness of $t \ge s$ implies $\mathbb{E}[L_s h(\zeta) K_s|\mathscr{F}_0] = \mathbb{E}[L_t h(\zeta) K_s|\mathscr{F}_0]$. Since h and K_s are also arbitrary, it follows that L is an $\mathbb{F}^{\sigma(\zeta)}$-martingale. Thus, for each $t \ge 0$ define the measure $\mathbb{P}^*$ on $\mathscr{F}_t^{\sigma(\zeta)}$ by $d\mathbb{P}^*|_{\mathscr{F}_t^{\sigma(\zeta)}} = L_t \; d\mathbb{P}|_{\mathscr{F}_t^{\sigma(\zeta)}}$. By Carathéodory's extension theorem the probability measure $\mathbb{P}^*$ can be uniquely extended to $\mathscr{F}_\infty^{\sigma(\zeta)}$. The equivalence of $\mathbb{P}^*$ and $\mathbb{P}$ on $\mathscr{F}_t^{\sigma(\zeta)}$ for each $t \in [0, \infty)$ follows from positivity of L_t. For any bounded Borel function h, any $\mathscr{F}_t$-measurable bounded random variable K_t, and denoting by $\mathbb{E}^*$ the expectation under $\mathbb{P}^*$, the above computations also yield

$$\mathbb{E}^*\left[h(\zeta) K_t|\mathscr{F}_0\right] = \mathbb{E}[h(\zeta)|\mathscr{F}_0]\, \mathbb{E}[K_t|\mathscr{F}_0]. \tag{4.20}$$

For $h = 1$ (resp. $K_t = 1$), one obtains $\mathbb{E}^*[K_t|\mathscr{F}_0] = \mathbb{E}[K_t|\mathscr{F}_0]$ (resp. $\mathbb{E}^*[h(\zeta)|\mathscr{F}_0] = \mathbb{E}[h(\zeta)|\mathscr{F}_0]$) and the assertions $\mathbb{P}^*|_{\mathscr{F}_t} = \mathbb{P}|_{\mathscr{F}_t}$ and $\mathbb{P}^*|_{\sigma(\zeta)} = \mathbb{P}|_{\sigma(\zeta)}$ are proven. Thus the identity (4.20) can be rewritten as $\mathbb{E}^*\left[h(\zeta) K_t|\mathscr{F}_0\right] = \mathbb{E}^*[h(\zeta)|\mathscr{F}_0]\, \mathbb{E}^*[K_t|\mathscr{F}_0]$ which shows that the random variable ζ and the σ-field $\mathscr{F}_t$ are conditionally independent given $\mathscr{F}_0$ under $\mathbb{P}^*$. Hence it follows that ζ and the σ-field $\mathscr{F}_t$ are also conditionally independent given $\mathscr{F}_0$ under $\mathbb{P}^*$. □

Remark 4.38 Note that one cannot claim that $\mathbb{P}^*$ is equivalent to $\mathbb{P}$ on $\mathscr{F}_\infty^{\sigma(\zeta)}$, since we do not know *a priori* whether $\frac{1}{p(\zeta)}$ is a *closed* $(\mathbb{F}^{\sigma(\zeta)}, \mathbb{P})$-martingale or not. A similar problem is studied by Föllmer and Imkeller in [95] (it is therein called a *paradox*) in the case where the reference (canonical) filtration is enlarged by means of the information about the endpoint at time $t = 1$. In our setting, this corresponds to the case where ζ is $\mathscr{F}_\infty$-measurable but not $\mathscr{F}_t$-measurable for any $t \in [0, \infty)$.

Corollary 4.39 *Under the probability measure* $\mathbb{P}^*$, $\mathbb{F}$ *is immersed in* $\mathbb{F}^{\sigma(\zeta)}$.

Proof Since under $\mathbb{P}^*$ the random variable ζ and the σ-field $\mathscr{F}_\infty$ are conditionally independent given $\mathscr{F}_0$ the assertion follows by Proposition 1.22. □

4.5.2 Hypothesis $\mathscr{H}'$ and Semimartingale Decomposition

In this subsection, we give a direct proof of the fact that any $\mathbb{F}$-local martingale X is a $\mathbb{F}^{\sigma(\zeta)}$-semimartingale. We also provide the $\mathbb{F}^{\sigma(\zeta)}$-canonical semimartingale decomposition of any $\mathbb{F}$-local martingale.

The following proposition is a particular case of the decomposition obtained under Jacod's absolute continuity condition in Theorem 4.25. The proof is easier but less general. The interest for us is that it shows the power of the change of probability measure discussed in Sect. 4.5.1.

Proposition 4.40 *Assume that ζ satisfies hypothesis $\mathscr{E}$ under $\mathbb{P}$. Then every $(\mathbb{F}, \mathbb{P})$-local martingale X is an $(\mathbb{F}^{\sigma(\zeta)}, \mathbb{P})$-semimartingale with canonical decomposition*

$$X_t = \widetilde{X}_t + \int_0^t \frac{1}{p_{s-}(\zeta)} d\langle X, p(u)\rangle_s^{\mathbb{F}}{}_{|u=\zeta},$$

where $\widetilde{X}$ is an $(\mathbb{F}^{\sigma(\zeta)}, \mathbb{P})$-local martingale.

Proof If X is an $(\mathbb{F}, \mathbb{P})$-martingale, it is also an $(\mathbb{F}^{\sigma(\zeta)}, \mathbb{P}^*)$-martingale. Indeed, since $\mathbb{P}$ and $\mathbb{P}^*$ are equal on $\mathbb{F}$, X is an $(\mathbb{F}, \mathbb{P}^*)$-martingale. Using the fact that, ζ and $\mathscr{F}_\infty$ are conditionally independent given $\mathscr{F}_0$ under $\mathbb{P}^*$, by Proposition 1.22, X is an $(\mathbb{F}^{\sigma(\zeta)}, \mathbb{P}^*)$-martingale. Recall from Remark 4.26 that, by [123, Theorem 2.5 (a)], for an $(\mathbb{F}, \mathbb{P})$-martingale X, there exists a η-negligible set B, such that $\langle X, p(u)\rangle^{\mathbb{F}}$ is well-defined for $u \notin B$. Then, recalling that $\frac{d\mathbb{P}}{d\mathbb{P}^*} = \frac{p_t(\zeta)}{p_0(\zeta)}$ on $\mathscr{F}_t^{\sigma(\zeta)}$, Girsanov's theorem implies that the process $\widetilde{X}$, defined by

$$\widetilde{X}_t = X_t - \int_0^t \frac{1}{p_{s-}(\zeta)} d\langle X, p(u)\rangle_s^{\mathbb{F}}{}_{|u=\zeta}$$

is an $(\mathbb{F}^{\sigma(\zeta)}, \mathbb{P})$-martingale. □

Remark 4.41 A random time τ is called $\mathscr{E}$-time if it satisfies hypothesis $\mathscr{E}$. Denote by $\mathbb{G}$ the progressive enlargement of $\mathbb{F}$ with τ. Note that $\mathbb{F} \subset \mathbb{G} \subset \mathbb{F}^{\sigma(\tau)}$ and that $\mathbb{F}$ and $\mathbb{G}$ coincide on $[\![0, \tau[\![$, and $\mathbb{G}$ and $\mathbb{F}^{\sigma(\tau)}$ coincide on $[\![\tau, \infty[\![$ (see Definition 1.28). Theorem 1.25 and Proposition 4.40 imply that any $(\mathbb{F}, \mathbb{P})$-local martingale is a $(\mathbb{G}, \mathbb{P})$-semimartingale. Proposition 1.24 and Corollary 4.39 imply that any $(\mathbb{F}, \mathbb{P}^*)$-local martingale is $(\mathbb{G}, \mathbb{P}^*)$-local martingale, in other words, $\mathbb{F}$ is immersed in $\mathbb{G}$ under $\mathbb{P}^*$.

4.5.3 Arbitrages

An important consequence of hypothesis $\mathscr{E}$ in finance is that it rules out arbitrage opportunities.

Lemma 4.42 *Assume that ζ satisfies hypothesis $\mathscr{E}$ under $\mathbb{P}$. Let Y be an $(\mathbb{F}, \mathbb{P})$-martingale. Then, the $\mathbb{F}^{\sigma(\zeta)}$-adapted process $\widetilde{Y} := YL$, where L is introduced in Theorem 4.37, is an $(\mathbb{F}^{\sigma(\zeta)}, \mathbb{P})$-martingale which satisfies $\mathbb{E}[\widetilde{Y}_t|\mathscr{F}_t] = Y_t$.*

Proof The process $\widetilde{Y}$ is an $(\mathbb{F}^{\sigma(\zeta)}, \mathbb{P})$-martingale if and only if for every $s \leq t$ and $A \in \mathscr{F}_s^{\sigma(\zeta)}$, one has $\mathbb{E}_{\mathbb{P}}[\widetilde{Y}_t \mathbb{1}_A] = \mathbb{E}_{\mathbb{P}}[\widetilde{Y}_s \mathbb{1}_A]$. By Theorem 4.37, this is equivalent to $\mathbb{E}^*[Y_t \mathbb{1}_A] = \mathbb{E}^*[Y_s \mathbb{1}_A]$. The last equality holds since the $(\mathbb{F}, \mathbb{P})$-martingale Y is also an $(\mathbb{F}, \mathbb{P}^*)$-martingale, as, again by Theorem 4.37, $\mathbb{P}$ and $\mathbb{P}^*$ coincide on $\mathbb{F}$. Therefore, the conditional independence given $\mathscr{F}_0$ of ζ and $\mathbb{F}$ under $\mathbb{P}^*$ implies, by Proposition 1.22, that Y is also an $(\mathbb{F}^{\sigma(\zeta)}, \mathbb{P}^*)$-martingale. Noting that $\mathbb{E}_{\mathbb{P}}[L_t|\mathscr{F}_t] = 1$ (take $s = t$ and $h = 1$ in (4.20)), we obtain

$$Y_t = \mathbb{E}^*[Y_t|\mathscr{F}_t] = \frac{\mathbb{E}_{\mathbb{P}}[L_t Y_t|\mathscr{F}_t]}{\mathbb{E}_{\mathbb{P}}[L_t|\mathscr{F}_t]} = \mathbb{E}_{\mathbb{P}}[\widetilde{Y}_t|\mathscr{F}_t]$$

and the result follows. □

Proposition 4.43 *Assume that ζ satisfies hypothesis $\mathscr{E}$ under $\mathbb{P}$.*
(a) Let $[0, T]$ be a finite time horizon. If $(S, \mathbb{F}, \mathbb{P})$ satisfies NFLVR, then $(S, \mathbb{F}^{\sigma(\zeta)}, \mathbb{P})$ satisfies NFLVR.
(b) If $(S, \mathbb{F}, \mathbb{P})$ admits a local martingale deflator Y (i.e., NUPBR in $\mathbb{F}$ is satisfied), then $(S, \mathbb{F}^{\sigma(\zeta)}, \mathbb{P})$ admits $Y/p(\zeta)$ as a local martingale deflator (i.e., NUPBR in $\mathbb{F}^{\sigma(\zeta)}$ is satisfied).

Proof (a) NFLVR is equivalent to the existence of a positive u.i. $(\mathbb{F}, \mathbb{P})$-martingale $(Y_t)_{t\in[0,T]}$ such that $(Y_t S_t)_{t\in[0,T]}$ is an $(\mathbb{F}, \mathbb{P})$-local martingale. It follows by Theorem 4.37 that $(Y_t)_{t\in[0,T]}$ is an $(\mathbb{F}, \mathbb{P}^*)$-martingale and $(Y_t S_t)_{t\in[0,T]}$ is an $(\mathbb{F}, \mathbb{P}^*)$-local martingale. Hence, by Corollary 4.39 (a), $(Y_t S_t)_{t\in[0,T]}$ is a $(\mathbb{F}^{\sigma(\zeta)}, \mathbb{P}^*)$-local martingale, and $\mathbb{Q}^*$, defined as $\frac{d\mathbb{Q}^*}{d\mathbb{P}}|_{\mathscr{F}_T^{\sigma(\zeta)}} = Y_T \frac{d\mathbb{P}^*}{d\mathbb{P}}|_{\mathscr{F}_T^{\sigma(\zeta)}}$ is an EMM in $\mathbb{F}^{\sigma(\zeta)}$ since $\mathbb{P}^*$ and $\mathbb{P}$ are equivalent on $\mathscr{F}_T$.
(b) By Lemma 4.42, the process $(Y_t L_t)_{t\in[0,\infty)}$ is an $\mathbb{F}^{\sigma(\zeta)}$-local martingale deflator for $(S_t)_{t\in[0,\infty)}$, where $L_t = p_0(\zeta)/p_t(\zeta)$ for $t < \infty$ and is introduced in Theorem 4.37. Therefore, since $p_0(\zeta)$ is positive $\mathbb{F}_0^{\sigma(\zeta)}$-measurable r.v., $(S_t)_{t\in[0,\infty)}$ admits $(Y_t/p_t(\zeta))_{t\in[0,\infty)}$ as an $\mathbb{F}^{\sigma(\zeta)}$-local martingale deflator. □

4.6 Some Examples in the Literature

We present now an incomplete list of examples, giving the references where the proofs can be found. Here, B is a Brownian motion, and B^* its running maximum, i.e., $B_t^* = \sup_{s\leq t}|B_s|$.

- $\zeta = A_\infty^{(-\mu)}$ where $A_t^{(\mu)} := \int_0^t e^{2(B_s+\mu s)}ds$. In that case, $\rho(x, t) = 2\mu - \frac{e^{2(B_t-\mu t)}}{x-A_t^{(-\mu)}}$. Mansuy and Yor [173, p. 34].

- $\zeta = (g_T, \mathrm{d}_T)$ where g_T is the last zero of B before T and d_T the first zero of B after T. Then $\rho((x,y),t) = -\frac{B_t}{x-t}1\!\!1_{\{y>x>t\}} + \left(\frac{1}{B_t} - \frac{B_t}{y-t}\right) 1\!\!1_{\{g_t<T\}}1\!\!1_{\{y>t\wedge T\}}1\!\!1_{\{x=g_t\}}$. Mansuy [171].
- $\zeta = \vartheta$ where $\vartheta = \sup\{t : B_t^* < 1\}$. If X is an $\mathbb{F}$-martingale satisfying the condition $\int_0^\vartheta \frac{1}{1-B_s}|d\langle X, B\rangle_s| < \infty$, then the process $X_t - \int_0^{t\wedge\vartheta} \frac{1}{1-B_s}\left(1 - \frac{(1-B_s)^2}{\vartheta-s}\right) d\langle X, B\rangle_s$ is an $\mathbb{F}^{\sigma(\vartheta)}$-local martingale. Jeulin [138, Lemma 3.25].
- $\zeta = A_\infty^p$ where A^p is the predictable part in the decomposition of the Azéma supermartingale Z associated with an honest time τ. Under conditions **(CA)**, $\rho(x,t) = \frac{1}{Z_t}1\!\!1_{\{x>A_t^p\}} - \frac{1}{1-Z_t}1\!\!1_{\{x=A_t^p\}}$. Nikeghbali [183, Theorem 9.9].
- $\zeta = T_1$ where T_1 is the first hitting time of 1 by a Brownian motion. Then, for $t \le T_1$, one has $\rho(x,t) = -\frac{1}{1-B_t} + \frac{1-B_t}{x-t}$. Mansuy and Yor [173, p. 34].
- $\zeta = (B_T, B_T^*)$, where B^* is the running maximum. Mansuy [171], Mansuy and Yor [173, p. 34].
- $\zeta = \ell_T$ where ℓ is the local time at level 0 of a Brownian motion. Yor [219, Chap. 12]
- $\zeta = (B_T, \ell_T)$ where ℓ is the local time at level 0 of a Brownian motion. Mansuy [171].
- $\zeta = L_T$ for an integrable Lévy process L, $L_t = \widetilde{L}_t + \int_0^t \frac{L_T - L_s}{T-s} ds$, where $\widetilde{L}$ is a martingale in the enlarged filtration. Jacod and Protter [124], Mansuy [171], Corcuera and Valdivia [66].
- $\zeta = T_n$ where T_n is the nth jump of a Poisson process. Corcuera and Valdivia [66].
- $\zeta = \alpha B_T + N_T$ where N is a Poisson process. Draouil and Øksendal [83].

Here are two examples where the initial enlargement is done with a σ-field generated by a process. In these cases, the method we have presented is not immediately applicable (mainly because the added σ-field is not generated by a random variable).

- $\mathscr{G}_t = \mathscr{F}_t \vee \mathscr{L}$ where $\mathscr{L}$ is the σ-field generated by the local time $(\ell_t, t \ge 0)$. Then, setting $\mathrm{d}_u = \inf\{t > u : B_t = 0\}$, it can be proved that $B_t = \widetilde{B}_t + \int_0^t (\frac{1}{B_s} - \frac{B_s}{\mathrm{d}_s - s}) ds$. Yor [219, Chap. 12]
- The filtration of the Brownian motion B is enlarged initially by the filtration generated by the increasing process B^*. Then, if X is an $\mathbb{F}$-martingale and if $\int_0^t \frac{1}{B_s^* - B_s} d\langle X, B\rangle_s < \infty$, the process $X_t + \int_0^t \frac{1}{B_s^* - B_s}\left(1 - \frac{B_s^* - B_s}{C_s - s}\right) d\langle B, X\rangle_s$ is a martingale in the enlarged filtration, where $C_s := \inf\{t > s : B_t^* > B_s^*\}$. See Jeulin [138, Lemma 3.25].

4.7 Bibliographic Notes

The study of initial enlargement of filtration started in 1976 when Itô [121] established that if B is a Brownian motion in its natural filtration, then B remains a semimartingale in the enlarged filtration $\mathbb{F}^{\sigma(B_1)} := \mathbb{F} \nabla \sigma(B_1)$. Itô obtained the decomposition of B as an $\mathbb{F}^{\sigma(B_1)}$-semimartingale, and he showed that one has the formula

$$B_1 \int_0^t H_s dB_s = \int_0^t B_1 H_s dB_s ,$$

where, on the right-hand side, the stochastic integral is with respect to $\mathbb{F}^{\sigma(B_1)}$-semimartingale B. This decomposition is recalled in Sect. 4.1.1, where we discuss Brownian bridge. Details on the filtration generated by a Brownian bridge are presented in Yor [219]. The reader will find interesting studies of Brownian bridges and associated filtrations in Brody et al. [46], Bedini [30] and Mansuy and Yor [174]. In [211], Wu defines a class of drifts such that the solution of $dX_t = dB_t + f(t, \widetilde{B}, X)dt$ is a Brownian motion and applies this study to insider trading in a Kyle–Back model in Föllmer et al. [97]. Poisson bridges are studied in Corcuera and Valdivia [66], Elliott et al. [89], Gasbarra et al. [104] and Jeulin and Yor [142]. Galtchouk [102] has extended the study (in particular the semimartingale decomposition formula) to any process with independent increments. Lévy bridges are studied in Jacod and Protter [124], Mansuy [171] and Mansuy and Yor [172]. Generalizations to dynamic Markov bridges with applications are developed in Campi et al. [50–52].

Yor's method, presented in Yor [216, 218] and generalized in Mansuy [171], strongly relies on the predictable representation property of the reference filtration. We have not covered a case of the weak representation property, e.g., representation w.r.t. a Poisson measure, for that, see Ankirchner [18] and Ankirchner and Zwierz [21].

Assuming the PRP in $\mathbb{F}$, its propagation to the enlarged filtration $\mathbb{G}$ is established under Jacod's equivalence condition in Amendinger [13, Theorem 2.4], Callegaro et al. [48], and under Jacod's absolute continuity condition in Fontana [99]. An important notion, especially in finance, is the multiplicity of a filtration, i.e., the minimal number of martingales which allow the PRP. A typical consequence of enlargement of filtration is an increase of the multiplicity. See Davis and Varaiya [69], Duffie [84] and Jeanblanc and Song [135].

The first papers on hypothesis $\mathscr{H}'$ in the initial enlargement of filtration setting are Jacod [123] and Jeulin [138, Chap. 3]. We recommand also Meyer [177]. In particular, an initial enlargement of filtration with atomic σ-field is discussed in Jeulin [138, Theorem 3.2]. Jacod's equivalence condition, appearing in the literature also under the name Hypothesis $\mathscr{E}$, is presented in a general setting in Amendinger [13, 14] and plays a fundamental role in Grorud and Pontier [110]. See also the paper by Danilova et al. [68]. Another presentation of hypothesis $\mathscr{E}$ is done in Ankirchner et al. [20] and Imkeller and Perkowski [119]. The right-continuity of the filtration $\mathbb{F} \vee \sigma(\zeta)$ is shown in Amendinger [13, Proposition 1.10] under Jacod's equivalence condition, and extended to Jacod's absolutely continuous condition in Fontana [99]. Li and Rutkowski [169] study the case where $\mathbb{P}(\tau \leq u | \mathscr{F}_t) = \int_{[0,u]} m_t(s) dD_s$ where D is decreasing and $m(s)$ is a parameterized positive martingale. They called these times τ pseudo-initial times. Choulli and Deng [58] study the stability of structure condition for an initial enlargement of filtration.

To the best of our knowledge, Karatzas and Pikovski [150] were the first to make use of enlargement of filtration for insider trading. This methodology was extended in

Grorud and Pontier [108, 110]. A statistical test to detect insider trading is presented in Grorud and Pontier [109].

Malliavin calculus techniques are presented in Corcuera et al. [65], Di Nunno et al. [80], Draouil and Øksendal [83], Imkeller [118], Imkeller et al. [120], Kohatsu-Higa [156], Kohatsu-Higa and Sulem [157, 158], León et al. [167] and Rindisbacher [191] to study the insider trading problem. Insider trading is sometimes studied via filtering conditions in the case where the dynamics of the prices are a diffusion, in which the drift is not measurable with respect to the driving Brownian motion. See Monoyios [178] and Fouque et al. [101]. Few papers are devoted to insider trading in the case where the prices are discontinuous semimartingales. Poisson bridges are used for modelling insider trading in Elliott and Jeanblanc [88], Gasbara et al. [104], Grorud and Pontier [108] and Kohatsu-Higa and Yamazoto [160]. The study of information drift is presented in details in the thesis of Ankirchner [17] and related papers by Ankirchner et al. [19], Amendinger et al. [16] and Amendinger et al. [15]. See Hillairet [114] and Hillairet and Jiao [116] for applications of pricing methodologies in various filtrations.

The study of enlargement with $\int f(s)dB_s$ (see Example 4.16) can be found in Chaleyat-Maurel and Yor [54], Nikeghbali [183] and Yor [219]. This is applied in mathematical finance in order to model default times in Crépey et al. [67]. See also Baudoin [29]. The enlargement with $\sup_{s\leq 1} B_s$ is presented in Jeulin [138, Proposition 3.24]. The enlargement with the supremum of a martingale is part of Mansuy's thesis [171].

NUPBR is studied in Acciaio et al. [2] and Aksamit et al. [10]. Imkeller and Perkowski [119] prove that for locally bounded processes, the NUPBR is equivalent to the existence of a dominating local martingale measure. Chau et al. [55] present applications.

4.8 Exercises

Exercise 4.1 In the setting of Sect. 4.2.2 verify that the set of Radon–Nikodym densities of $\mathbb{F}^{\sigma(N_T)}$-EMM's given in (4.4) is not empty.

Exercise 4.2 Find the $\mathbb{F}^{\sigma(\zeta)}$-semimartingale decomposition of $\mathbb{F}$-martingales when $\mathbb{F}$ is a Brownian filtration and $\zeta = 1\!\!1_{\{a\leq B_T\leq b\}}$. Describe the arbitrage opportunities in $\mathbb{F}^{\sigma(\zeta)}$.

Chapter 5
Progressive Enlargement

We study the progressive enlargement of the reference filtration $\mathbb{F} := (\mathscr{F}_t, t \geq 0)$, given on a probability space $(\Omega, \mathscr{F}, \mathbb{P})$ and satisfying the usual conditions, with a random time τ. Recall that a random time is a $[0, \infty]$-valued random variable. For a random time τ, $\mathbb{A} := (\mathscr{A}_t, t \geq 0)$ is the natural filtration of the default indicator process $A := 1\!\!1_{[\![\tau,\infty[\![}$. We define, as in Sect. 2.2, two progressively enlarged filtrations $\mathbb{G}^0 := (\mathscr{G}_t^0)_{t\geq 0}$ and $\mathbb{G} := (\mathscr{G}_t)_{t\geq 0}$ by

$$\mathscr{G}_t^0 := \mathscr{F}_t \vee \mathscr{A}_t \quad \text{and} \quad \mathscr{G}_t := \cap_{\varepsilon>0}\mathscr{G}_{t+\varepsilon}^0 ,$$

or, equivalently $\mathbb{G}^0 := \mathbb{F} \vee \mathbb{A}$ and $\mathbb{G} := \mathbb{F} \triangledown \mathbb{A}$. In this chapter:

- We present the $\mathbb{G}$-semimartingale decomposition of $\mathbb{F}$-martingales stopped at τ, valid for any random time τ.
- We study four classes of random times: pseudo-stopping times, honest times, $\mathscr{J}$-times and thin times.
- We exhibit arbitrage opportunities which may occur in the enlarged filtration and we give a sufficient condition for NUPBR.

Let us recall the notation introduced in Chap. 1, Sect. 1.3.4: $Z = {}^o(1 - A)$, $\widetilde{Z} = {}^o(1 - A_-)$ and $Z = m - A^o = n - A^p$, where m and n are $\mathbb{F}$-martingales and A^o (resp. A^p) is the $\mathbb{F}$-dual optional (resp. predictable) projection of A.

5.1 $\mathbb{G}$-Semimartingale Decomposition of $\mathbb{F}$-Martingales Stopped at a Random Time τ

We prove that, for every càdlàg $\mathbb{F}$-local martingale X, the stopped process X^τ is a $\mathbb{G}$-semimartingale. In Theorems 5.1 and 5.2, useful $\mathbb{G}$-semimartingale decompositions of X^τ are presented.

A. Aksamit and M. Jeanblanc, *Enlargement of Filtration with Finance in View*, SpringerBriefs in Quantitative Finance,
https://doi.org/10.1007/978-3-319-41255-9_5

Theorem 5.1 *Every càdlàg $\mathbb{F}$-local martingale X stopped at time τ is a special $\mathbb{G}$-semimartingale with the canonical decomposition*

$$X_t^\tau = \widehat{X}_t + \int_0^{t\wedge\tau} \frac{d\langle X, m\rangle_s^{\mathbb{F}}}{Z_{s-}} = \widehat{X}_t + \int_0^{t\wedge\tau} \frac{d\langle X, n\rangle_s^{\mathbb{F}} + dJ_s}{Z_{s-}} \tag{5.1}$$

where $\widehat{X}$ is a $\mathbb{G}$-local martingale and J is the $\mathbb{F}$-dual predictable projection of the process $\mathbb{1}_{[\![\tau,\infty[\![}\Delta X_\tau$.

Proof Note that since any $\mathbb{F}$-local martingale X is locally in $H^1(\mathbb{F})$, by Proposition 1.49 (a), the processes $[X, m]$ and $[X, n]$ have locally integrable variation thus $\langle X, m\rangle^{\mathbb{F}}$ and $\langle X, n\rangle^{\mathbb{F}}$ exist. Let X be an $\mathbb{F}$-martingale and H be a bounded $\mathbb{G}$-predictable process with the $(\mathbb{F}, \tau)$-predictable reduction h (see Proposition 2.11). Hence, the process h is bounded and $\mathbb{F}$-predictable such that $H\mathbb{1}_{[\![0,\tau]\!]} = h\mathbb{1}_{[\![0,\tau]\!]}$ and $h\mathbb{1}_{\{Z_-=0\}} = 0$. We therefore have

$$\begin{aligned}
\mathbb{E}\left[(H\cdot X^\tau)_\infty\right] &= \mathbb{E}\left[(h\cdot X)_\tau\right] = \mathbb{E}\left[[h\cdot X, m]_\infty\right] = \mathbb{E}\left[h\cdot [X, m]_\infty\right] \\
&= \mathbb{E}\left[(h\cdot \langle X, m\rangle^{\mathbb{F}})_\infty\right] = \mathbb{E}\left[\int_0^\infty \frac{h_s Z_{s-}}{Z_{s-}} \mathbb{1}_{\{Z_{s-}>0\}} d\langle X, m\rangle_s^{\mathbb{F}}\right] \\
&= \mathbb{E}\left[\int_0^\tau \frac{h_s}{Z_{s-}} d\langle X, m\rangle_s^{\mathbb{F}}\right] = \mathbb{E}\left[\int_0^\tau \frac{H_s}{Z_{s-}} d\langle X, m\rangle_s^{\mathbb{F}}\right],
\end{aligned}$$

where the second equality comes from Proposition 1.49 (b), and the second to last one from equalities (1.9) and (1.8). By considering $\mathbb{G}$-predictable processes of the form $\mathbb{1}_G\mathbb{1}_{]\!]s,t]\!]}$ for $s \le t$ and $G \in \mathcal{G}_s$ we conclude the first equality in (5.1) for $\mathbb{F}$-martingales. Since $m = n + A^o - A^p$, by Proposition 1.49 (b) and Proposition 1.16, we obtain

$$\begin{aligned}
\mathbb{E}\left[h\cdot [X, m]_\infty\right] &= \mathbb{E}\left[h\cdot [X, n]_\infty\right] + \mathbb{E}\left[h\cdot [X, A^o - A^p]_\infty\right] \\
&= \mathbb{E}\left[h\cdot \langle X, n\rangle_\infty^{\mathbb{F}}\right] + \mathbb{E}\left[(h\Delta X\cdot A^o)_\infty\right] + \mathbb{E}\left[(h\Delta A^p\cdot X)_\infty\right] \\
&= \mathbb{E}\left[\int_0^\tau \frac{h_s}{Z_{s-}} d\langle X, n\rangle_s^{\mathbb{F}}\right] + \mathbb{E}\left[\left(h\cdot \left(\mathbb{1}_{[\![\tau,\infty[\![}\Delta X_\tau\right)\right)_\infty\right] \\
&= \mathbb{E}\left[\int_0^\tau \frac{h_s}{Z_{s-}} d\langle X, n\rangle_s^{\mathbb{F}}\right] + \mathbb{E}\left[\int_0^\tau \frac{h_s}{Z_{s-}} d\left(\mathbb{1}_{[\![\tau,\infty[\![}\Delta X_\tau\right)_s^{p,\mathbb{F}}\right],
\end{aligned}$$

where in the second equality we have used the fact that $h\Delta A^p\cdot X$ is an $\mathbb{F}$-martingale as A^p has bounded jumps. Therefore, as for the first equality in (5.1), we conclude that the process $X^\tau - \int_0^{\cdot\wedge\tau} \frac{d\langle X, n\rangle_s^{\mathbb{F}} + dJ_s}{Z_{s-}}$ is a $\mathbb{G}$-martingale. Moreover it is equal to $\widehat{X}$ from the uniqueness of the canonical decomposition of a special semimartingale. The result is extended to $\mathbb{F}$-local martingales by localization. □

We now give another $\mathbb{G}$-semimartingale decomposition of an $\mathbb{F}$-local martingale X stopped at τ, where the finite variation part is no longer a predictable process. Note that this decomposition is a particular optional decomposition.

Theorem 5.2 *Every càdlàg $\mathbb{F}$-local martingale X stopped at time τ is a $\mathbb{G}$-semimartingale with decomposition*

$$X_t^\tau = \bar{X}_t + \int_0^{t\wedge\tau} \frac{1}{\widetilde{Z}_s} d[X,m]_s - \left(\mathbb{1}_{[\![\widetilde{R},\infty[\![}\, \Delta X_{\widetilde{R}}\right)^{p,\mathbb{F}}_{t\wedge\tau} \tag{5.2}$$

where $\bar{X}$ is a $\mathbb{G}$-local martingale and $\widetilde{R} := R_{\{\widetilde{Z}_R=0<Z_{R-}\}}$, where R, defined in Lemma 2.14, is the first time when Z vanishes, i.e., $R := \inf\{t : Z_t = 0\}$.

Proof Assume that X is an $\mathbb{F}$-martingale. Let H be a bounded $\mathbb{G}$-predictable process and h its $(\mathbb{F},\tau)$-predictable reduction (see Proposition 2.11). Using Proposition 1.49 (b) and the fact that $\widetilde{Z}$ is the optional projection of $\mathbb{1}_{[\![0,\tau]\!]}$, we deduce that

$$\begin{aligned}
\mathbb{E}\left[(H\cdot X^\tau)_\infty\right] &= \mathbb{E}\left[[h\cdot X, m]_\infty\right] \\
&= \mathbb{E}\left[\int_0^\infty \frac{h_s \widetilde{Z}_s}{\widetilde{Z}_s} \mathbb{1}_{\{\widetilde{Z}_s>0\}} d[X,m]_s\right] + \mathbb{E}\left[\int_0^\infty h_s \mathbb{1}_{\{\widetilde{Z}_s=0<Z_{s-}\}} d[X,m]_s\right] \\
&= \mathbb{E}\left[\int_0^\tau \frac{h_s}{\widetilde{Z}_s} \mathbb{1}_{\{\widetilde{Z}_s>0\}} d[X,m]_s\right] - \mathbb{E}\left[\int_0^\infty h_s Z_{s-} d\left(\Delta X_{\widetilde{R}}\, \mathbb{1}_{[\![\widetilde{R},\infty[\![}\right)_s\right],
\end{aligned}$$

where we have also made use of the fact that, since $\{\widetilde{Z} = 0 < Z_-\} = [\![\widetilde{R}]\!]$,

$$\mathbb{1}_{\{\widetilde{Z}=0<Z_-\}} \cdot [X,m] = \Delta X_{\widetilde{R}}\, \Delta m_{\widetilde{R}}\, \mathbb{1}_{[\![\widetilde{R},\infty[\![} = -Z_{\widetilde{R}-} \Delta X_{\widetilde{R}}\, \mathbb{1}_{[\![\widetilde{R},\infty[\![}$$

as, by Proposition 1.46 (d), $\Delta m = \widetilde{Z} - Z_-$. Then, since by Lemma 2.14 $\widetilde{Z} > 0$ on $[\![0,\tau]\!]$, we find

$$\begin{aligned}
\mathbb{E}\left[(H\cdot X^\tau)_\infty\right] &= \mathbb{E}\left[\int_0^\tau \frac{h_s}{\widetilde{Z}_s} d[X,m]_s\right] - \mathbb{E}\left[\int_0^\infty h_s Z_{s-} d\left(\Delta X_{\widetilde{R}} \mathbb{1}_{[\![\widetilde{R},\infty[\![}\right)^{p,\mathbb{F}}_s\right] \\
&= \mathbb{E}\left[\int_0^\tau \frac{H_s}{\widetilde{Z}_s} d[X,m]_s\right] - \mathbb{E}\left[\int_0^\tau H_s d\left(\Delta X_{\widetilde{R}} \mathbb{1}_{[\![\widetilde{R},\infty[\![}\right)^{p,\mathbb{F}}_s\right].
\end{aligned}$$

By considering $\mathbb{G}$-predictable processes of the form $\mathbb{1}_G \mathbb{1}_{]\!]s,t]\!]}$ for $s \le t$ and $G \in \mathscr{G}_s$ we conclude that $\bar{X}$ is a $\mathbb{G}$-martingale and the result follows by localization. □

Remark 5.3 (a) Under condition (**C**), both decompositions given in (5.1) and (5.2) coincide since $\widetilde{R} = \infty$ (it is a predictable jump of the martingale m), $[X,m] = \langle X,m\rangle^{\mathbb{F}}$ (as $[X,m]$ is continuous) and $\widetilde{Z} = Z_-$ (by Proposition 1.46 (d)).
(b) Let us introduce the filtration $\widetilde{\mathbb{G}} := (\widetilde{\mathscr{G}}_t)_{t\ge 0}$ where

$$\widetilde{\mathscr{G}}_t := \{F \in \mathscr{F}_\infty \vee \sigma(\tau) : \exists\, \widetilde{F} \in \mathscr{F}_t \text{ s.t. } F \cap \{\tau > t\} = \widetilde{F} \cap \{\tau > t\}\}.$$

First of all, note that each $\widetilde{\mathscr{G}}_t$ is indeed a σ-field. One can also show that

$$\widetilde{\mathscr{G}}_t = \mathscr{F}_t \vee \sigma\left(\widetilde{A}^F_s,\ s \le t,\ F \in \mathscr{F}_\infty\right) \quad \text{where} \quad \widetilde{A}^F = 1\!\!1_F 1\!\!1_{[\![\tau,\infty[\![}.$$

Thus, one has $\mathbb{F} \subset \mathbb{G}^0 \subset \widetilde{\mathbb{G}}$. Moreover, all the three filtrations $\mathbb{G}^0$, $\mathbb{G}$ and $\widetilde{\mathbb{G}}$ coincide on $[\![0,\tau[\![$ (see Definition 1.28 for the notion of coincidence of filtrations). Furthermore, the filtration $\widetilde{\mathbb{G}}$ coincides on $[\![\tau,\infty[\![$ with the constant filtration $\mathbb{G}^\infty := (\mathscr{G}_\infty)_{t\ge 0}$ where $\mathscr{G}_\infty = \mathscr{F}_\infty \vee \sigma(\tau)$. Then, by Lemma 2.9, for any bounded $\mathscr{F}$-measurable r.v. Y,

$$\mathbb{E}[Y|\widetilde{\mathscr{G}}_t] = 1\!\!1_{\{t<\tau\}} \frac{\mathbb{E}[Y 1\!\!1_{\{t<\tau\}}|\mathscr{F}_t]}{Z_t} + 1\!\!1_{\{\tau \le t\}} \mathbb{E}[Y|\mathscr{G}_\infty]. \tag{5.3}$$

By Theorem 1.10, the right-hand side in (5.3) has a right-continuous modification. Then, by Theorem 1.11, we have the following equalities

$$\mathbb{E}[Y|\widetilde{\mathscr{G}}_{t+}] = \lim_{u\downarrow t} \mathbb{E}[Y|\widetilde{\mathscr{G}}_u] = \mathbb{E}[Y|\widetilde{\mathscr{G}}_t].$$

Thus, by taking $Y = 1\!\!1_{\widetilde{G}}$ with $\widetilde{G} \in \widetilde{\mathscr{G}}_{t+}$, we conclude that $\widetilde{\mathbb{G}}$ is a right-continuous filtration. Furthermore, by Proposition 1.29, any $\mathbb{G}$-local martingale stopped at τ is a $\widetilde{\mathbb{G}}$-local martingale and vice versa. It is immediate that, for any càdlàg $\mathbb{F}$-local martingale X stopped at time τ, the formulas (5.1) and (5.2) are valid in any filtration between $\mathbb{G}$ and $\widetilde{\mathbb{G}}$ (or even $\mathbb{G}^0$ and $\widetilde{\mathbb{G}}$).

5.2 Honest Times

5.2.1 Definition and Properties

Definition 5.4 *A random time τ is an $\mathbb{F}$-honest time if, for every $t > 0$, there exists an $\mathscr{F}_t$-measurable random variable τ_t such that $\tau = \tau_t$ on $\{\tau < t\}$. Without loss of generality, we assume that $\tau_t \le t$.*

Note that a finite valued $\mathbb{F}$-honest time is $\mathscr{F}_\infty$-measurable. The above definition can be modified as follows.

Proposition 5.5 *(a) A random time τ is an $\mathbb{F}$-honest time if and only if, for every $t > 0$, there exists an $\mathscr{F}_{t-}$-measurable r.v. τ_t such that $\tau = \tau_t$ on $\{\tau < t\}$.*
(b) A random time τ is an $\mathbb{F}$-honest time if and only if, for every $t > 0$, there exists an $\mathscr{F}_t$-measurable r.v. τ_t such that $\tau = \tau_t$ on $\{\tau \le t\}$.

Proof In both cases only necessity has to be shown. Let τ be an honest time. Thus there exists an $\mathscr{F}_t$-measurable r.v. such that $\tau = \tau_t$ on $\{\tau < t\}$ for each $t > 0$. Let us define the process α^- as $\alpha^-_t = \sup_{r\in\mathbb{Q}^+, r<t} \tau_r$. This definition implies that α^- is an $\mathbb{F}$-adapted process with increasing, left-continuous paths such that $\alpha^-_t = \tau$ on $\{\tau < t\}$. Since α^- is $\mathbb{F}$-predictable, the assertion (a) is proven.

Denote by α the right-limit of α^-, i.e., $\alpha_t = \alpha^-_{t+}$. Then, α is an $\mathbb{F}$-adapted increasing process such that $\alpha_t = \tau$ on $\{\tau \le t\}$ and (b) follows. □

The following corollary is a refinement of (2.5) for honest times.

Corollary 5.6 *Let τ be an $\mathbb{F}$-honest time. For a fixed $t \ge 0$, a r.v. Y is $\mathcal{G}^0_t$-measurable if and only if $Y = y 1\!\!1_{\{t<\tau\}} + \widetilde{y} 1\!\!1_{\{\tau \le t\}}$ for some $\mathcal{F}_t$-measurable r.v.'s y and $\widetilde{y}$.*

Proof By Proposition 2.8, $Y = y 1\!\!1_{\{t<\tau\}} + \widehat{y}(\tau) 1\!\!1_{\{\tau \le t\}}$ for an $\mathcal{F}_t$-measurable random variable y and an $\mathcal{F}_t \otimes \mathcal{B}(\mathbb{R}^+)$-measurable mapping $(\omega, u) \to \widehat{y}(\omega, u)$. Hence, applying Proposition 5.5 (b), it is enough to take $\widetilde{y}(\omega) := \widehat{y}(\omega, \tau_t(\omega))$ which is clearly $\mathcal{F}_t$-measurable. □

Example 5.7 (a) Any $\mathbb{F}$-stopping time T is an $\mathbb{F}$-honest time since $T = T \wedge t$ on $\{T < t\}$ and $T \wedge t$ is $\mathcal{F}_t$-measurable.
(b) Let $\mathbb{F}^B$ be the natural filtration of a Brownian motion B and define, for any t, the random time $g_t = \sup\{s < t : B_s = 0\}$ and the stopping time $T_1 := \inf\{s : B_s = 1\}$. The random time g_{T_1} is an $\mathbb{F}^B$-honest time since, for $t > 0$, we have $g_{T_1} = g_t$ on $\{g_{T_1} < t\}$, and g_t is $\mathcal{F}^B_t$-measurable.
(c) Let X be a continuous process and $\mathbb{F}^X$ its natural filtration. Assume that X^*_∞ exists where $X^*_t := \sup_{s \le t} X_s$. The random time $\vartheta := \sup\{s : X_s = X^*_\infty\}$ is an $\mathbb{F}^X$-honest time since, on the set $\{\vartheta < t\}$, one has $\vartheta = \sup\{s : s \le t,\ X_s = X^*_t\}$.

In the following theorem, we collect useful equivalent characterisations of the honest time property.

Theorem 5.8 *A random time τ is an $\mathbb{F}$-honest time if and only if one of the following equivalent conditions holds:*
(a) there exists an $\mathbb{F}$-optional set Γ such that $\tau(\omega) = \sup\{t : (t, \omega) \in \Gamma\}$ on $\{\tau < \infty\}$, with the convention that $\sup \emptyset = \infty$;
(b) $\widetilde{Z}_\tau = 1$ on $\{\tau < \infty\}$;
(c) $\tau = \sup\{t : \widetilde{Z}_t = 1\}$ on $\{\tau < \infty\}$;
(d) $A^o_\cdot = A^o_{\cdot \wedge \tau}$;
(e) any $\mathbb{G}$-predictable process Y is decomposable as $Y = y 1\!\!1_{[\![0,\tau]\!]} + \widetilde{y} 1\!\!1_{]\!]\tau,\infty[\![}$ where y and $\widetilde{y}$ are two $\mathbb{F}$-predictable processes.

Proof Suppose that τ is an $\mathbb{F}$-honest time and let α be the $\mathbb{F}$-optional process defined in the proof of Proposition 5.5. Then $\tau = \sup\{t : \alpha_t = t\}$ on the set $\{\tau < \infty\}$. Note that the quantity $\sup\{t : \alpha_t = t\}$ is finite on the set $\{\tau < \infty\}$. Thus, τ being an honest time implies (a).

Assume (a). Let γ be the end of the optional set Γ and $\widetilde{Z}^\gamma$ the associated Azéma supermartingale. We necessarily have that $\gamma \le \tau$ and $\widetilde{Z}^\gamma \le \widetilde{Z}$. Then, by Proposition 1.54 (or Lemma 1.53), we obtain $\Gamma \subset \{\widetilde{Z}^\gamma = 1\} \subset [\![0, \gamma]\!]$. Thus, on $\{\tau < \infty\}$, we have $1 \ge \widetilde{Z}_\tau \ge \widetilde{Z}^\gamma_\tau = \widetilde{Z}^\gamma_\gamma = 1$ as also $\tau = \gamma$ on $\{\tau < \infty\}$, which implies (b).

As $0 = \mathbb{E}[1\!\!1_{\{\widetilde{Z}_t=1\}}(1 - \widetilde{Z}_t)] = \mathbb{E}[1\!\!1_{\{\widetilde{Z}_t=1\}} 1\!\!1_{\{\tau<t\}}]$ we obtain the equivalence between (b) and (c).

Condition (c) trivially implies that τ is an honest time, since we can simply take $\tau_t := \sup\{s \le t : \widetilde{Z}_s = 1\}$ which is $\mathscr{F}_t$-measurable and $\tau = \tau_t$ on $\{\tau < t\}$.

Assume (c). Thus, we have that $[\![\tau]\!] \subset \{\widetilde{Z} = 1\}$ where $\{\widetilde{Z} = 1\}$ is an $\mathbb{F}$-optional set. By Proposition 1.54 (or Lemma 1.52), we obtain that the support of dA^o is the smallest optional set containing $[\![\tau]\!]$, hence it must be included in $\{\widetilde{Z} = 1\}$. Then (d) follows from (c). If (d) holds, then on $\{\tau < t\}$, τ equals to $\inf\{s : s \le t, A^o_s = A^o_t\}$ which is $\mathscr{F}_t$-measurable, hence τ is an honest time.

By Proposition 2.11 (b), $Y = \mathbb{1}_{[\![0,\tau]\!]} y + \mathbb{1}_{]\!]\tau,\infty[\![} \, \widehat{y}(\tau)$ where y is $\mathbb{F}$-predictable and $(\omega, t, u) \mapsto \widehat{y}_t(\omega, u)$ is a $\mathscr{P}(\mathbb{F}) \otimes \mathscr{B}(\mathbb{R}^+)$-measurable function. And, if τ is honest, then (e) holds since it is enough to take $\widetilde{y}_t(\omega) := \widehat{y}(\omega, t, \alpha^-_t(\omega))$ with α^- introduced in the proof of Proposition 5.5, which is clearly $\mathbb{F}$-predictable.

If (e) holds, then looking at $Y_t = \tau \mathbb{1}_{\{t > \tau\}}$, the honesty of τ is obtained. □

Remark 5.9 We emphasize that in Definition 5.4, Proposition 5.5 and Theorem 5.8 we require equalities to hold for each $\omega \in \Omega$. However, since we always assume that $\mathbb{F}$ satisfies the usual conditions, it is equivalent to require almost sure equalities. Note that in an a.s. formulation, the same assertions hold. There is some weak dependence of honest times on a probability measure $\mathbb{P}$, only through the $\mathbb{P}$-null sets.

Lemma 5.10 *If τ is an $\mathbb{F}$-honest time, then $\mathbb{G} = \mathbb{G}^0 = \mathbb{G}^*$ where $\mathbb{G}^* := (\mathscr{G}^*_t, t \ge 0)$*

$$\mathscr{G}^*_t := \{F \in \mathscr{F} : F = (\widehat{F} \cap \{\tau \le t\}) \cup (\widetilde{F} \cap \{\tau > t\}) \text{ for some } \widehat{F}, \widetilde{F} \in \mathscr{F}_t\}.$$

Proof Note that $\mathscr{G}^*_t$ is indeed a σ-field for each t. To show that $\mathscr{G}^*_t \subset \mathscr{G}^*_u$ for each $u \ge t$ let us consider $F = (\widehat{F} \cap \{\tau \le t\}) \cup (\widetilde{F} \cap \{\tau > t\})$ with $\widehat{F}, \widetilde{F} \in \mathscr{F}_t$. Then

$$F = (\widehat{F} \cap \{\tau \le u\} \cap \{\tau_u \le t\}) \cup (\widetilde{F} \cap \{\tau \le u\} \cap \{\tau_u > t\}) \cup (\widetilde{F} \cap \{\tau > u\})$$

where τ_u is an $\mathscr{F}_u$-measurable r.v. such that $\tau = \tau_u$ on $\{\tau \le u\}$ (Proposition 5.5 (b)). Thus F is of the desired form and belongs to $\mathscr{G}^*_u$. Similarly, right-continuity of $\mathbb{F}$ implies right-continuity of $\mathbb{G}^*$.

The sets of the form $\{\tau \le s\}$ with $s \le t$ belong to $\mathscr{G}^*_t$:

$$\{\tau \le s\} = (\{\tau_t \le s\} \cap \{\tau \le t\}) \cup (\emptyset \cap \{\tau > t\}) \in \mathscr{G}^*_t.$$

Thus, by the monotone class theorem that $\mathbb{G}^0 \subset \mathbb{G}^*$. Since the inclusion $\mathbb{G}^* \subset \mathbb{G}^0$ is straightforward, the equality $\mathbb{G}^* = \mathbb{G}^0$ follows. □

Lemma 2.9 extends in the case of honest times to the following result.

Lemma 5.11 *Assume that τ is an $\mathbb{F}$-honest time. Let X be an $\mathscr{F}$-measurable integrable r.v. Then, for any $t \ge 0$,*

$$\mathbb{E}[X|\mathscr{G}_t] = \frac{1}{Z_t}\mathbb{E}\left[X\mathbb{1}_{\{t<\tau\}}|\mathscr{F}_t\right]\mathbb{1}_{\{t<\tau\}} + \frac{1}{1-Z_t}\mathbb{E}\left[X\mathbb{1}_{\{t\ge\tau\}}|\mathscr{F}_t\right]\mathbb{1}_{\{t\ge\tau\}}. \tag{5.4}$$

Proof (a) The form of the first term on the right-hand side follows by Lemma 2.9. The second term is well-defined as $0 = \mathbb{E}[\mathbb{1}_{\{Z_t=1\}}(1 - Z_t)] = \mathbb{E}[\mathbb{1}_{\{Z_t=1\}}\mathbb{1}_{\{\tau<t\}}]$ and its form follows by the same arguments as in the proof of Lemma 2.9, taking into account Theorem 5.8 (e) and the fact that $\mathbb{G} = \mathbb{G}^0$ (as shown in Lemma 5.10). □

Corollary 5.12 *Let τ be an $\mathbb{F}$-honest time. Then every $\mathbb{G}$-optional process Y admits a decomposition*

$$Y = L\mathbb{1}_{[\![0,\tau[\![} + J\mathbb{1}_{[\![\tau]\!]} + K\mathbb{1}_{]\!]\tau,\infty[\![},$$

where L and K are $\mathbb{F}$-optional processes and where J is an $\mathbb{F}$-progressively measurable process.

Proof Similarly as in Proposition 2.11, it is enough to establish the decomposition for any u.i. càdlàg $\mathbb{G}$-martingale X. Also similarly as in Proposition 2.11, we will aggregate the decomposition from Lemma 5.11 (which is given for each t). The first term on the right-hand of (5.4) does not pose a problem since, by Lemma 2.14, the random sets $\{Z = 0\}$ and $[\![0, \tau[\![$ are disjoint. Therefore the first term is right-continuous and $\mathbb{F}$-adapted on $[\![0, \tau[\![$. The second term in (5.4) does not pose a problem when restricted to $]\!]\tau, \infty[\![$ since $\{Z = 1\} \subset \{\tilde{Z} = 1\} \subset [\![0, \tau]\!]$, where the last inclusion is due to Theorem 5.8 (c) and thus the second term is right-continuous and $\mathbb{F}$-adapted on $]\!]\tau, \infty[\![$. Therefore it remains to deal with $[\![\tau]\!] \cap \{Z = 1\}$ which is a thin set on which the càdlàg $\mathbb{G}$-martingale X is defined by its right-limit. Thus the assertion holds. □

Example 5.13 We recall Barlow's counterexample given in [28, p. 319] to show that a $\mathbb{G}$-optional process cannot always be decomposed as $L\mathbb{1}_{[\![0,\tau[\![} + K\mathbb{1}_{[\![\tau,\infty[\![}$, where L and K are $\mathbb{F}$-optional processes. Let B be a Brownian motion, $\mathbb{F}$ its natural filtration $\vartheta = \inf\{t : |B_t| = 1\}$, $\tau = \sup\{t \le \vartheta : B_t = 0\}$ and $\mathbb{G}$ the progressive enlargement of $\mathbb{F}$ with τ. The process X defined as $X_t = \mathbb{1}_{\{t\ge\tau\}}\mathrm{sgn}(B_\vartheta)$ is right-continuous and $\mathbb{G}$-adapted, hence $\mathbb{G}$-optional. Moreover X is a $\mathbb{G}$-martingale. Obviously, if the pair (L, K) exists, then $L = 0$ and one can choose K to be $\mathbb{F}$-predictable, since $\mathcal{O}(\mathbb{F}) = \mathcal{P}(\mathbb{F})$. Then $\Delta X_\tau = K_\tau$ would be $\mathcal{G}_{\tau-}$-measurable, which contradicts the $\mathbb{G}$-martingale property of X.

The following lemma is given in [28, Lemma 3.1].

Lemma 5.14 *Let Y be a càdlàg $\mathbb{G}$-adapted process such that Y_t is integrable for each t and let $A := \mathbb{1}_{[\![\tau,\infty[\![}$. Then Y is a $\mathbb{G}$-martingale if and only if the following conditions are satisfied:*
(a) for $s \le t$, $\mathbb{E}[Y_t|\mathcal{F}_s] = \mathbb{E}[Y_s|\mathcal{F}_s]$,
(b) for $s \le t$, $\mathbb{E}[A_sY_t|\mathcal{F}_s] = \mathbb{E}[A_sY_s|\mathcal{F}_s]$.

Proof By Lemma 5.11, the process Y is a $\mathbb{G}$-martingale if and only if for each $s \le t$, one has $\mathbb{E}\left[(Y_t - Y_s)\mathbb{1}_{\{s<\tau\}}|\mathcal{F}_s\right] = 0$ and $\mathbb{E}\left[(Y_t - Y_s)\mathbb{1}_{\{s\ge\tau\}}|\mathcal{F}_s\right] = 0$. □

5.2.2 $\mathbb{G}$-Semimartingale Decomposition of an $\mathbb{F}$-Martingale

Proposition 5.15 *Let τ be an $\mathbb{F}$-honest time. Every càdlàg $\mathbb{F}$-local martingale X is a special $\mathbb{G}$-semimartingale with canonical decomposition*

$$X_t = \widetilde{X}_t + \int_0^{t\wedge\tau} \frac{1}{Z_{s-}} d\langle X, m\rangle_s^{\mathbb{F}} - \int_\tau^{\tau\vee t} \frac{1}{1-Z_{s-}} d\langle X, m\rangle_s^{\mathbb{F}}, \tag{5.5}$$

where $\widetilde{X}$ is a $\mathbb{G}$-local martingale.

Proof The part up to τ was obtained in Theorem 5.1. The proof for the part after τ is analogous, taking into account Theorem 5.8 (e). □

5.2.3 Multiplicative Decomposition of Z Under (CA)

Theorem 5.16 *Let τ be a finite $\mathbb{F}$-honest time and Z its Azéma supermartingale. Assume conditions **(CA)**. Then, there exists a continuous and non-negative $\mathbb{F}$-local martingale N, with $N_0 = 1$ and $\lim_{t\to\infty} N_t = 0$, s.t. $Z = \frac{N}{N^*}$ where $N_t^* := \sup_{s\le t} N_s$. Moreover, every càdàg $\mathbb{F}$-local martingale X is a special $\mathbb{G}$-semimartingale with canonical decomposition given by*

$$X_t = \widetilde{X}_t + \int_0^{t\wedge\tau} \frac{1}{N_s} d\langle X, N\rangle_s^{\mathbb{F}} - \int_\tau^{\tau\vee t} \frac{1}{N_s^* - N_s} d\langle X, N\rangle_s^{\mathbb{F}}.$$

Proof Under condition **(A)**, by Proposition 1.46 (d) and Lemma 1.48 (a), $Z = \widetilde{Z}$. Then, since τ is an honest time, by Theorem 5.8 (b), $Z_\tau = 1$ on $\{\tau < \infty\}$, thus $[\![\tau]\!] \subset \{Z = 1\}$ where $\{Z = 1\}$ is a predictable set (Z is continuous under conditions **(CA)**). By Proposition 1.54, we obtain that the support of the measure dA^p is the smallest predictable set which contains $[\![\tau]\!]$, hence it is included in $\{Z = 1\}$. We now prove that the process $N := Z\exp(A^p)$ is an $\mathbb{F}$-local martingale. Indeed, since $Z = n - A^p$ (see Proposition 1.46 (a)),

$$\begin{aligned} Z_t \exp(A_t^p) &= 1 + \int_0^t \exp(A_s^p)\, dZ_s + \int_0^t Z_s \exp(A_s^p) dA_s^p \\ &= 1 + \int_0^t \exp(A_s^p)\, dn_s + \int_0^t (Z_s - 1)\exp(A_s^p) dA_s^p = 1 + \int_0^t \exp(A_s^p)\, dn_s. \end{aligned}$$

Since A^p is continuous, increasing with support in $\{Z = 1\}$, one has $N^* = \exp(A^p)$, thus $Z = \frac{N}{N^*}$. Since, by Theorem 5.8 (c) and condition **(A)**, $\tau = \sup\{t : Z_t = 1\}$ we conclude that $\tau = \sup\{t : N_t = N_t^*\}$. Moreover as $Z_0 = 1$ and $Z_\infty = 0$ we obtain $N_0 = 1$ and $N_\infty := \lim_{t\to\infty} N_t = 0$. Furthermore, relating the processes N and N^* with the Doob–Meyer decomposition of Z we obtain that

$$n_t = 1 + \frac{1}{N^*} \cdot N_t = \mathbb{E}[\log N^*_\infty | \mathscr{F}_t],$$

where the last equality follows from the definition of n given in Proposition 1.46. The decomposition (5.5) of an $\mathbb{F}$-local martingale X can be thus written as

$$X_t = \widetilde{X}_t + \int_0^{t\wedge\tau} \frac{1}{N_s} d\langle X, N\rangle^{\mathbb{F}}_s - \int_\tau^{\tau\vee t} \frac{1}{N^*_s - N_s} d\langle X, N\rangle^{\mathbb{F}}_s .$$

□

Remark 5.17 Note that the $\mathbb{G}$-semimartingale decomposition from Theorem 5.16 is the same as the one for initial enlargement with N^*_∞, presented in (4.6), Example 4.12. This is due to the fact that the drift term which appears in the formula for the initial enlargement is $\mathbb{G}$-predictable.

5.2.4 Examples of Honest Times

We present some examples of honest times and we give their associated Azéma's supermartingales.

• Last Passage Time of a Transient Diffusion

Lemma 5.18 *Let τ be the last passage time of an $\mathbb{F}$-adapted process X below a deterministic level a, i.e., $\tau = \sup\{t : X_t \le a\}$. Then, τ is an $\mathbb{F}$-honest time.*

Proof On the set $\{\tau < u\}$, we have that $\tau = \tau_u$ for $\tau_u = \sup\{t \le u : X_t \le a\}$ and τ_u is $\mathscr{F}_u$-measurable. □

We show that, in a diffusion set-up, the Doob–Meyer decomposition of the Azéma supermartingale Z may be computed explicitly for some last passage times. Denote by $\mathbb{P}_x$ the probability measure under which $\mathbb{P}_x(X_0 = x) = 1$. Recall that a scale function s is an non-decreasing function from $\mathbb{R}$ to $\mathbb{R}$ such that, for $x \in [a, b]$

$$\mathbb{P}_x(T_a < T_b) = \frac{s(x) - s(b)}{s(a) - s(b)}$$

where $T_y = \inf\{t : X_t = y\}$. One chooses a scale function such that $s(\infty) = 0$ and hence, $s(y) < 0$ for $y \in \mathbb{R}$. See [190, Chap. VII, Sect. 3] for the further properties of a scale function.

Proposition 5.19 *Let X be a homogeneous diffusion such that $X_t \to \infty$ when $t \to \infty$, and s be a scale function of X such that $s(\infty) = 0$. Define $\tau_a := \sup\{t : X_t = a\}$. Then,*

$$\mathbb{P}_x(\tau_a > t | \mathscr{F}_t) = \frac{s(X_t)}{s(a)} \wedge 1 .$$

The $\mathbb{F}^X$-dual predictable projection of the random time τ_a equals

$$A^p = -\frac{1}{2s(a)} L^{s(a)}(Y),$$

where $L^{s(a)}(Y)$ is the local time process of the martingale $Y = s(X)$ at level $s(a)$.

Proof Observe that

$$\begin{aligned}\mathbb{P}_x(\tau_a > t|\mathscr{F}_t) &= \mathbb{P}_x\left(\inf_{u \geq t} X_u < a \,\Big|\, \mathscr{F}_t\right) = \mathbb{P}_x\left(\sup_{u \geq t}(-s(X_u)) > -s(a) \,\Big|\, \mathscr{F}_t\right) \\ &= \mathbb{P}_{X_t}\left(\sup_{u \geq 0}(-s(X_u)) > -s(a)\right) = \frac{s(X_t)}{s(a)} \wedge 1,\end{aligned}$$

where we have used the Markov property of X and Doob's maximal identity presented in Exercise 1.6. The form of A^p comes from $y \wedge 1 = y - (y-1)^+$ and the Itô–Tanaka formula (see, e.g., [136, Sect. 4.1.8]). It follows that

$$\frac{s(X_t)}{s(a)} \wedge 1 = \mu_t + \frac{1}{2s(a)} L_t^{s(a)}(Y),$$

where μ is a martingale. Recall that $s(a) < 0$, so the process $\frac{1}{2s(a)} L^{s(a)}(Y)$ is decreasing. The required result follows. □

• Last Passage Time at 0 of a Brownian Motion Before a Fixed Time

Let $\mathbb{F}$ be the natural filtration of a Brownian motion B and g_T be the last passage time at level 0 of B before a fixed finite time T, defined as

$$g_T := \sup\{t \leq T : B_t = 0\}.$$

Note that $g_T \leq T$.

Proposition 5.20 *The random time g_T is an $\mathbb{F}$-honest time; its $\mathbb{F}$-dual predictable projection A^p and its Azéma's supermartingale Z are given by*

$$A_t^p = \sqrt{\frac{2}{\pi}} \int_0^{t \wedge T} \frac{dL_s}{\sqrt{T-s}} \quad \text{and} \quad Z_t = \Phi\left(\frac{|B_t|}{\sqrt{T-t}}\right) \mathbb{1}_{\{t<T\}},$$

where L is the local time of B at 0 and $\Phi(x) = \sqrt{\frac{2}{\pi}} \int_x^\infty e^{-u^2/2} du$.

Proof Let $t < T$. The set $\{g_T > t\}$ is equal to $\{d_t < T\}$ with d_t defined by

$$d_t := \inf\{u \geq t : B_u = 0\} = t + \inf\{u \geq 0 : B_{t+u} - B_t = -B_t\},$$

where $(B_{t+u} - B_t, u \geq 0)$ is a Brownian motion independent of $\mathscr{F}_t$. Thus, by [136, Proposition 4.3.3.1], the conditional law of d_t given $\mathscr{F}_t$ is equal to the law of random variable $t + \frac{y^2}{G^2}$, evaluated at $y = B_t$, where G is a standard Gaussian variable. From

$$\mathbb{P}\left(\frac{y^2}{G^2} < T - t\right) = \Phi\left(\frac{|y|}{\sqrt{T-t}}\right) \quad \text{where} \quad \Phi(x) = \sqrt{\frac{2}{\pi}} \int_x^\infty e^{-u^2/2} du\,,$$

we deduce that, for $t < T$,

$$Z_t := \mathbb{P}(g_T > t | \mathscr{F}_t) = \mathbb{P}(d_t < T | \mathscr{F}_t) = \Phi\left(\frac{|B_t|}{\sqrt{T-t}}\right).$$

From the Itô–Tanaka formula, we obtain that

$$d\frac{|B_s|}{\sqrt{T-s}} = \frac{1}{\sqrt{T-s}} \operatorname{sgn}(B_s) dB_s + \frac{1}{\sqrt{T-s}} dL_s + \frac{|B_s|}{2(T-s)^{3/2}} ds.$$

The Itô formula, the fact that L increases only on the set $\{t : B_t = 0\}$ and the identity $x\Phi'(x) + \Phi''(x) = 0$ lead to

$$\begin{aligned}
Z_t &= \int_0^{t\wedge T} \Phi'\left(\frac{|B_s|}{\sqrt{T-s}}\right) d\frac{|B_s|}{\sqrt{T-s}} + \frac{1}{2}\int_0^{t\wedge T} \Phi''\left(\frac{|B_s|}{\sqrt{T-s}}\right) \frac{ds}{T-s} \\
&= \int_0^{t\wedge T} \Phi'\left(\frac{|B_s|}{\sqrt{T-s}}\right) \frac{\operatorname{sgn}(B_s)}{\sqrt{T-s}} dB_s + \int_0^{t\wedge T} \Phi'\left(\frac{|B_s|}{\sqrt{T-s}}\right) \frac{dL_s}{\sqrt{T-s}} \\
&= \int_0^{t\wedge T} \Phi'\left(\frac{|B_s|}{\sqrt{T-s}}\right) \frac{\operatorname{sgn}(B_s)}{\sqrt{T-s}} dB_s - \sqrt{\frac{2}{\pi}} \int_0^{t\wedge T} \frac{dL_s}{\sqrt{T-s}}.
\end{aligned}$$

The result follows. □

Corollary 5.21 *The decomposition of the Brownian motion B in the progressive enlargement $\mathbb{G}$ of the filtration $\mathbb{F}$ with g_T is therefore given by*

$$B_t = \widetilde{B}_t + \int_0^{t\wedge g_T} \frac{\Phi'}{\Phi}\left(\frac{|B_s|}{\sqrt{T-s}}\right) \frac{\operatorname{sgn}(B_s)}{\sqrt{T-s}} ds - \int_{g_T}^{(t\vee g_T)\wedge T} \frac{\Phi'}{1-\Phi}\left(\frac{|B_s|}{\sqrt{T-s}}\right) \frac{\operatorname{sgn}(B_s)}{\sqrt{T-s}} ds\,,$$

where $\widetilde{B}$ is a $\mathbb{G}$-Brownian motion.

• Last Passage Time of a Brownian Motion Before Hitting a Level

Let $X_t = x + \sigma B_t$ where x and σ are positive constants and B a Brownian motion with natural filtration $\mathbb{F}$. We consider, for $0 < a < x$, the last passage time of X at the level a before hitting the level 0, given as

$$g^a_{T_0}(X) := \sup\{t \leq T_0 : X_t = a\} \quad \text{where} \quad T_0 := T_0(X) := \inf\{t \geq 0 : X_t = 0\}.$$

Proposition 5.22 *The random time $g^a_{T_0}$ is an $\mathbb{F}$-honest time; its $\mathbb{F}$-dual predictable projection A^p and its Azéma's supermartingale Z are given by*

$$A^p_t = \frac{1}{2a} L^a_{t\wedge T_0(X)}(X) = \frac{\sigma}{2a} L^\alpha_{t\wedge T_0(B)}(B) \quad and \quad Z_t = \frac{X_{t\wedge T_0}}{a} \wedge 1\,,$$

where $L^\alpha(B)$ (resp. $L^a(X)$) is the local time of the Brownian Motion B (resp. of the diffusion X) at level $\alpha := \frac{a-x}{\sigma}$ (resp. level a).

Proof Set $T_y(B) := \inf\{t : B_t = y\}$ and $d^\alpha_t(B) := \inf\{s \geq t : B_s = \alpha\}$. Then $T_{-\frac{x}{\sigma}}(B) = T_0(X)$ and $\mathbb{P}\big(g^a_{T_0}(X) > t|\mathscr{F}_t\big) = \mathbb{P}\big(d^\alpha_t(B) < T_0|\mathscr{F}_t\big)$. Using computations done in [136, Proposition 3.5.1.1], it can be proven that

$$\mathbb{P}\big(g^a_{T_0}(X) > t|\mathscr{F}_t\big) = \frac{X_{t\wedge T_0}}{a} \wedge 1.$$

As a consequence, using $y \wedge 1 = 1 - (1-y)^+$ and the Itô–Tanaka formula, we obtain

$$Z_t = 1 + \frac{1}{a}\int_0^{t\wedge T_0} \mathbb{1}_{\{X_s<a\}} dX_s - \frac{1}{2a} L^a_{t\wedge T_0}(X) = 1 + \frac{\sigma}{a}\int_0^{t\wedge T_0} \mathbb{1}_{\{B_s<\alpha\}} dB_s - \frac{\sigma}{2a} L^\alpha_{t\wedge T_0}(B).$$

The result follows by Proposition 1.46 (a). □

Corollary 5.23 *The decomposition of the Brownian motion B in the progressive enlargement $\mathbb{G}$ of the filtration $\mathbb{F}$ with $g^a_{T_0}$ is therefore given by*

$$B_t = \widetilde{B}_t + \int_0^{t\wedge g^a_{T_0}} \frac{\sigma}{X_s} \mathbb{1}_{\{X_s<a\}} ds - \int_{g^a_{T_0}}^{g^a_{T_0}\vee t} \frac{\sigma}{a - X_s} \mathbb{1}_{\{X_s<a\}} \mathbb{1}_{\{s\leq T_0\}} ds.$$

5.3 $\mathscr{J}$-Times

The random time τ is called a **$\mathscr{J}$-time** if it satisfies Jacod's absolute continuity condition from Definition 4.13. A special case of $\mathscr{J}$-times is the class of $\mathscr{E}$-times which satisfy Jacod's equivalent condition (also called hypothesis $\mathscr{E}$) and are mentioned in Remark 4.41. See Sects. 4.4 and 4.5 for related results and notation.

5.3.1 Projections and Martingales

Lemma 2.9 extends in the case of $\mathscr{J}$-times to the following result.

Lemma 5.24 *Let τ be a $\mathscr{J}$-time, $T \in \mathbb{R}^+$ and the function $(\omega, u) \to X(\omega, u)$ be $\mathscr{F}_T \otimes \mathscr{B}(\overline{\mathbb{R}}^+)$-measurable, and either non-negative or bounded. Then, for $t \leq T$,*

$$\mathbb{E}[X(\tau)|\mathcal{G}_t] = \frac{\mathbb{1}_{\{t<\tau\}}}{Z_t}\mathbb{E}\left[\int_{(t,\infty]} X(u)p_T(u)\eta(du)\Big|\mathcal{F}_t\right] + \frac{\mathbb{1}_{\{\tau\le t\}}}{p_t(\tau)}\mathbb{E}\left[X(u)p_T(u)|\mathcal{F}_t\right]_{|u=\tau}$$

and, for $t > T$,

$$\mathbb{E}[X(\tau)|\mathcal{G}_t] = \mathbb{1}_{\{t<\tau\}}\frac{1}{Z_t}\int_{(t,\infty]} X(u)p_t(u)\eta(du) + \mathbb{1}_{\{\tau\le t\}}X(\tau).$$

Proof Let $t \le T$. By Lemma 2.9 and Proposition 4.18 applied to $Y_s(\omega,u) := X(u)\mathbb{1}_{\{t<u\}}\mathbb{1}_{\{s\ge T\}}$, we have

$$\begin{aligned}\mathbb{1}_{\{t<\tau\}}\mathbb{E}[X(\tau)|\mathcal{G}_t] &= \mathbb{1}_{\{t<\tau\}}\frac{\mathbb{E}\left[\mathbb{E}[X(\tau)\mathbb{1}_{\{t<\tau\}}|\mathcal{F}_T]|\mathcal{F}_t\right]}{Z_t}\\ &= \mathbb{1}_{\{t<\tau\}}\frac{1}{Z_t}\mathbb{E}\left[\int_{(t,\infty]} X(u)p_T(u)\eta(du)\Big|\mathcal{F}_t\right].\end{aligned}$$

Furthermore, by Corollary 4.21, we deduce that

$$\mathbb{1}_{\{\tau\le t\}}\mathbb{E}[X(\tau)|\mathcal{G}_t] = \mathbb{1}_{\{\tau\le t\}}\mathbb{E}\left[\mathbb{E}\left[X(\tau)|\mathcal{F}_t^{\sigma(\tau)}\right]\Big|\mathcal{G}_t\right] = \frac{\mathbb{1}_{\{\tau\le t\}}}{p_t(\tau)}\mathbb{E}\left[X(u)p_T(u)|\mathcal{F}_t\right]_{|u=\tau}$$

which completes the proof for $t \le T$. The case $t > T$ is now straightforward. □

We now compute the $\mathbb{G}$-predictable (resp. $\mathbb{G}$-optional) projection of an $\mathbb{F}^{\sigma(\tau)}$-predictable (resp. $\mathbb{F}^{\sigma(\tau)}$-optional) process.

Proposition 5.25 *Let τ be a $\mathcal{J}$-time. The following assertions hold.*
(a) Let the function $(\omega,t,u) \to Y_t(\omega,u)$ be $\mathcal{P}(\mathbb{F})\otimes\mathcal{B}(\overline{\mathbb{R}}^+)$-measurable, and either non-negative or bounded. Then, the $\mathbb{G}$-predictable projection of the process $Y(\tau)$ is given by

$${}^{p,\mathbb{G}}(Y(\tau))_t = \mathbb{1}_{\{t\le\tau\}}\frac{\int_{(t,\infty]} Y_t(u)p_{t-}(u)\eta(du)}{Z_{t-}} + \mathbb{1}_{\{t>\tau\}}Y_t(\tau), \quad t\ge 0.$$

(b) Let the function $(\omega,t,u) \to Y_t(\omega,u)$ be $\mathcal{O}(\mathbb{F})\otimes\mathcal{B}(\overline{\mathbb{R}}^+)$-measurable, and either non-negative or bounded. Then, the $\mathbb{G}$-optional projection of the process $Y(\tau)$ is given by

$${}^{o,\mathbb{G}}(Y(\tau))_t = \mathbb{1}_{\{t<\tau\}}\frac{\int_{(t,\infty]} Y_t(u)p_t(u)\eta(du)}{Z_t} + \mathbb{1}_{\{t\ge\tau\}}Y_t(\tau), \quad t\ge 0.$$

Proof (a) Let us first note that

$${}^{p,\mathbb{G}}(Y(\tau)) = {}^{p,\mathbb{G}}(Y(\tau)\mathbb{1}_{[\![0,\tau]\!]}) + {}^{p,\mathbb{G}}(Y(\tau)\mathbb{1}_{]\!]\tau,\infty[\![}).$$

Using Definition 1.28, the filtrations $\mathbb{F}^{\sigma(\tau)}$ and $\mathbb{G}$ coincide on the $\mathbb{G}$-predictable interval $]\!]\tau,\infty[\![$. The process $1\!\!1_{]\!]\tau,\infty[\![}Y(\tau)$, being an $\mathbb{F}^{\sigma(\tau)}$-predictable bounded process (see Proposition 4.22 (c)), is also $\mathbb{G}$-predictable. Thus ${}^{p,\mathbb{G}}(Y(\tau)1\!\!1_{]\!]\tau,\infty[\![}) = Y(\tau)1\!\!1_{]\!]\tau,\infty[\![}$. Moreover Propositions 2.11 (b) and 4.18 (a) imply that

$$ {}^{p,\mathbb{G}}(Y(\tau)1\!\!1_{[\![0,\tau]\!]})_t = 1\!\!1_{\{t\le\tau\}}\frac{1}{Z_{t-}}\,{}^{p,\mathbb{F}}(Y(\tau)1\!\!1_{[\![0,\tau]\!]})_t = \frac{1\!\!1_{\{t\le\tau\}}}{Z_{t-}}\int_t^\infty Y_t(u)p_{t-}(u)\eta(du) $$

which completes the proof of (a).

The proof of (b) follows by analogous arguments. □

Proposition 5.26 *Let τ be a $\mathcal{J}$-time. The Azéma supermartingales Z and $\widetilde{Z}$*

$$ Z_t = \int_{(t,\infty]} p_t(u)\eta(du) \quad \text{and} \quad \widetilde{Z}_t = \int_{[t,\infty]} p_t(u)\eta(du) $$

admit the following predictable and optional decompositions

$$ Z_t = n_t - \int_{[0,t]} p_{u-}(u)\,\eta(du) = m_t - \int_{[0,t]} p_u(u)\,\eta(du);\ \ \widetilde{Z}_t = m_t - \int_{[0,t)} p_u(u)\,\eta(du) $$

where n and m are $\mathbb{F}$-martingales given by

$$ n_t := 1 - \int_{[0,t]} (p_t(u) - p_{u-}(u))\,\eta(du) \quad \text{and} \quad m_t := 1 - \int_{[0,t]} (p_t(u) - p_u(u))\,\eta(du). $$

Proof Deriving expressions for Z, $\widetilde{Z}$ and their decompositions is straightforward. It remains to show that n and m are indeed $\mathbb{F}$-martingales. For each $u \in \overline{\mathbb{R}}^+$, $p(u)$ is an $\mathbb{F}$-martingale and hence, in a similar way to Example 1.33, for $s < u$, $\mathbb{E}[\Delta p_u(u)|\mathcal{F}_s] = 0$ since $\mathbb{E}[\Delta p_u(u)|\mathcal{F}_{u-}] = 0$ and $\mathcal{F}_s \subset \mathcal{F}_{u-}$. Therefore, for $s \le t$, $\mathbb{E}[n_s - n_t|\mathcal{F}_s]$ equals

$$ \begin{aligned} &= \mathbb{E}\left[\int_{[0,s]} (p_t(u) - p_s(u))\eta(du)\Big|\mathcal{F}_s\right] + \mathbb{E}\left[\int_{(s,t]} (p_t(u) - p_{u-}(u))\eta(du)\Big|\mathcal{F}_s\right] \\ &= \int_{(s,t]} \mathbb{E}[p_t(u) - p_{u-}(u)|\mathcal{F}_s]\eta(du) = \int_{(s,t]} \mathbb{E}[\Delta p_u(u)|\mathcal{F}_s]\eta(du) = 0 \end{aligned} $$

which completes the proof. □

Corollary 5.27 *Let τ be a $\mathcal{J}$-time and m, n be the martingales associated to the Azéma supermartingales. Then:*

(a) The martingale m is continuous and the jump of n equals $\Delta n_t = -\Delta p_t(t)\eta(\{t\})$.
(b) The $\mathbb{F}$-dual predictable projection of τ is $A^p_t = \int_{[0,t]} p_{u-}(u)\eta(du)$ and the $\mathbb{F}$-dual optional projection of τ is $A^o_t = \int_{[0,t]} p_u(u)\eta(du)$. Thus, we deduce from Proposition 2.15 that the process M defined by

$$M_t := A_t - \int_0^{t\wedge\tau} \frac{p_{u-}(u)}{Z_{u-}}\eta(du), \quad t \geq 0,$$

is a $\mathbb{G}$*-martingale.*

Proposition 5.28 *Assume that* τ *is a* $\mathcal{J}$*-time. Then,* $\mathbb{F}$ *is immersed in* $\mathbb{G}$ *if and only if for each* t*,* $p_t(u) = p_u(u)$ *for* η*-a.e.* u *satisfying* $u < t$*.*

Proof To show the necessary condition, we use Theorem 3.2 (c) (see also Lemma 3.8). If $\mathbb{F} \hookrightarrow \mathbb{G}$ and $h : \overline{\mathbb{R}}^+ \to \mathbb{R}$ is a bounded Borel function, then

$$\mathbb{E}[h(\tau)1\!\!1_{\{\tau\leq u\}}|\mathcal{F}_u] = \mathbb{E}[h(\tau)1\!\!1_{\{\tau\leq u\}}|\mathcal{F}_t] \quad \text{for all} \quad t > u.$$

Since τ is a $\mathcal{J}$-time, this leads to

$$\int_{[0,u]} h(s)p_u(s)\eta(ds) = \int_{[0,u]} h(s)p_t(s)\eta(ds) \quad \text{for all} \quad t > u.$$

Therefore the assertion. To show the sufficient condition we use the same identities. □

5.3.2 $\mathbb{G}$-Martingales Versus $\mathbb{F}$-Martingales

The goal of this subsection is to characterize $\mathbb{G}$-martingales in terms of $\mathbb{F}$-martingales.

Proposition 5.29 *A* $\mathbb{G}$*-optional process of the form* $Y := \widetilde{y}1\!\!1_{[\![0,\tau[\![} + \widehat{y}(\tau)1\!\!1_{[\![\tau,\infty[\![}$*, where* $\widetilde{y}$ *is* $\mathbb{F}$*-optional and* $(\omega, t, u) \to \widehat{y}_t(\omega, u)$ *is* $\mathcal{O}(\mathbb{F}) \otimes \mathcal{B}(\overline{\mathbb{R}}^+)$*-measurable, is a* $\mathbb{G}$*-martingale if and only if the following two conditions are satisfied*
(a) for η*-a.e* u*,* $\big(\widehat{y}_t(u)p_t(u), t \geq u\big)$ *is an* $\mathbb{F}$*-martingale;*
(b) the process y *is an* $\mathbb{F}$*-martingale, where*

$$y_t := \mathbb{E}[Y_t|\mathcal{F}_t] = \widetilde{y}_t Z_t + \int_0^t \widehat{y}_t(u)p_t(u)\eta(du)\,. \tag{5.6}$$

Proof For the necessity, as a first step, we show that we can reduce our attention to the case where Y is u.i. Indeed, let Y be a $\mathbb{G}$-martingale and $(T_n)_{n\geq 0}$ be a $\mathbb{G}$-localizing sequence such that, for each n, the associated stopped martingale $(Y_{t\wedge T_n}, t \geq 0)$ is u.i. Assuming that the result is established for u.i. martingales will prove that the processes in (a) and (b) are martingales up to T_n for each n. Since $T_n \to \infty$ as $n \to \infty$, the result follows.

Let Y be a u.i. $\mathbb{G}$-martingale. By Propositions 1.24 (b) and 4.33, $Y_t = \mathbb{E}[Y_t(\tau)|\mathcal{G}_t]$ for an $\mathcal{O}(\mathbb{F}) \otimes \mathcal{B}(\overline{\mathbb{R}}^+)$-measurable mapping $(\omega, t, u) \to Y_t(\omega, u)$ such that for η-almost every u the process $\big(Y_t(u)p_t(u), t \geq u\big)$ is an $\mathbb{F}$-martingale. One also has that

$$\mathbb{1}_{\{\tau\le t\}}\widehat{y}_t(\tau) = \mathbb{1}_{\{\tau\le t\}}Y_t = \mathbb{1}_{\{\tau\le t\}}\mathbb{E}[Y_t(\tau)|\mathscr{G}_t] = \mathbb{1}_{\{\tau\le t\}}Y_t(\tau)\,,$$

which implies, when combined with Lemma 4.23, that for η-almost every $u \le t$, the identity $Y_t(u) = \widehat{y}_t(u)$ holds $\mathbb{P}$-a.s. So, (a) is shown. Moreover, since Y is a $\mathbb{G}$-martingale, its $\mathbb{F}$-optional projection, namely the process y in (5.6), is an $\mathbb{F}$-martingale.

Conversely, assuming (a) and (b), we shall verify that $\mathbb{E}[Y_t|\mathscr{G}_s] = Y_s$ for $s \le t$. Let us first note that, by Lemma 2.9,

$$\mathbb{E}[Y_t|\mathscr{G}_s] = \mathbb{1}_{\{s<\tau\}}\frac{1}{Z_s}\mathbb{E}[Y_t\mathbb{1}_{\{s<\tau\}}|\mathscr{F}_s] + \mathbb{1}_{\{\tau\le s\}}\mathbb{E}[Y_t\mathbb{1}_{\{\tau\le s\}}|\mathscr{G}_s]\,. \tag{5.7}$$

We then compute the two conditional expectations on the right-hand side of (5.7):

$$\begin{aligned}
\mathbb{E}[Y_t\mathbb{1}_{\{s<\tau\}}|\mathscr{F}_s] &= \mathbb{E}[Y_t|\mathscr{F}_s] - \mathbb{E}[Y_t\mathbb{1}_{\{\tau\le s\}}|\mathscr{F}_s]\\
&= \mathbb{E}[y_t|\mathscr{F}_s] - \mathbb{E}\left[\mathbb{E}[\widehat{y}_t(\tau)\mathbb{1}_{\{\tau\le s\}}|\mathscr{F}_t]|\mathscr{F}_s\right]\\
&= y_s - \mathbb{E}\left[\int_{[0,s]}\widehat{y}_t(u)p_t(u)\eta(du)\Big|\mathscr{F}_s\right]\\
&= \widetilde{y}_s Z_s + \int_{[0,s]}\widehat{y}_s(u)p_s(u)\eta(du) - \int_{[0,s]}\widehat{y}_s(u)p_s(u)\eta(du) = \widetilde{y}_s Z_s
\end{aligned}$$

where we have used Fubini's theorem and the condition (a) to obtain the next-to-last identity. For the second term, an application of Lemma 5.24 yields

$$\begin{aligned}
\mathbb{E}[Y_t\mathbb{1}_{\{\tau\le s\}}|\mathscr{G}_s] &= \mathbb{E}[\widehat{y}_t(\tau)\mathbb{1}_{\{\tau\le s\}}|\mathscr{G}_s] = \mathbb{1}_{\{\tau\le s\}}\frac{1}{p_s(\tau)}\mathbb{E}\left[\widehat{y}_t(u)p_t(u)|\mathscr{F}_s\right]_{|u=\tau}\\
&= \mathbb{1}_{\{\tau\le s\}}\frac{1}{p_s(\tau)}\widehat{y}_s(\tau)p_s(\tau) = \mathbb{1}_{\{\tau\le s\}}\widehat{y}_s(\tau)
\end{aligned}$$

where the next-to-last identity holds in view of the condition (b). Thus the result. □

5.3.3 Hypothesis $\mathscr{H}'$ and Semimartingale Decomposition

Recall that, by Theorem 4.25, any $\mathbb{F}$-local martingale X is an $\mathbb{F}^{\sigma(\tau)}$-semimartingale. Since X is a $\mathbb{G}$-adapted process, in view of Stricker's Theorem 1.25, X is also a $\mathbb{G}$-semimartingale. The following proposition aims to obtain the $\mathbb{G}$-canonical decomposition of an $\mathbb{F}$-local martingale.

Proposition 5.30 *Let τ be a $\mathscr{J}$-time. Then, any càdlàg $\mathbb{F}$-local martingale X is a special $\mathbb{G}$-semimartingale with canonical decomposition*

$$X_t = \widehat{X}_t + \int_0^{t\wedge\tau}\frac{1}{Z_{s-}}d\langle X, m\rangle_s^{\mathbb{F}} + \int_{t\wedge\tau}^t \frac{1}{p_{s-}(\tau)}d\langle X, p(u)\rangle_s^{\mathbb{F}}{}_{|u=\tau}\,, \tag{5.8}$$

where $\widehat{X}$ is a $\mathbb{G}$-local martingale.

Proof The proof follows by Proposition 1.29 and Theorems 4.25 and 5.1. □

Remark 5.31 (a) Remark that $\mathscr{J}$-times are a second class of random times for which hypothesis $\mathscr{H}'$ is satisfied. The first class with this property was the class of honest times. In Sect. 5.4 we will discuss a third class (the one of thin times) which satisfies hypothesis $\mathscr{H}'$ (see Theorem 5.33).
(b) Note that honest times and $\mathscr{J}$-times exploit different properties. Namely, an honest time τ is a $\mathscr{J}$-time if and only if τ takes countably many values. Let τ be an honest time and τ_t the corresponding family of random variables (see Definition 5.4). Then for a bounded Borel function $f : \overline{\mathbb{R}}^+ \to \mathbb{R}$ we have

$$\begin{aligned}\mathbb{E}[f(\tau)|\mathscr{F}_t] &= \mathbb{E}[f(\tau)1\!\!1_{(\tau\le t)}|\mathscr{F}_t] + \mathbb{E}[f(\tau)1\!\!1_{(\tau>t)}|\mathscr{F}_t]\\ &= f(\tau_t)(1-Z_t) + \mathbb{E}\left[\int_t^\infty f(s)dA_s^o + f(\infty)Z_\infty\Big|\mathscr{F}_t\right].\end{aligned}$$

In particular, for $f(s) = 1\!\!1_{\{s>u\}}$, we get

$$\begin{aligned}\mathbb{P}(\tau > u|\mathscr{F}_t) &= 1\!\!1_{\{\tau_t>u\}}(1-Z_t) + \mathbb{E}\left[A_\infty^o - A_{t\vee u}^o + Z_\infty|\mathscr{F}_t\right]\\ &= \int_u^\infty (1-Z_t)\delta_{\tau_t}(ds) + \mathbb{E}\left[A_\infty^o - A_{t\vee u}^o + Z_\infty|\mathscr{F}_t\right].\end{aligned}$$

Assume that η, the law of τ, is not purely atomic. Denote by D the set of atoms of η and take t such that $\eta([0,t)\backslash D) > 0$. Then, the first term of conditional law $\delta_{\tau_t}1\!\!1_{\{\tau<t\}} = \delta_\tau 1\!\!1_{\{\tau<t\}}$ is not absolutely continuous with respect to η $\mathbb{P}$-a.s. as it is enough to take set $\{\tau \in [0,t)\backslash D\}$ which has positive probability.

5.4 Thin Times

In this section we present some results concerning thin times (see [9]). We are particularly interested in the hypothesis $\mathscr{H}'$. Recall that, from Definition 1.40, a thin time τ is built of $\mathbb{F}$-stopping times, i.e., $\tau = \infty 1\!\!1_{C_0} + \sum_n T_n 1\!\!1_{C_n}$ where $(T_n)_n$ is a sequence of $\mathbb{F}$-stopping times with disjoint graphs and

$$C_0 := \{\tau = \infty\} \quad \text{and} \quad C_n := \{\tau = T_n < \infty\} \quad \text{for} \quad n \ge 1. \tag{5.9}$$

We denote by z^n the càdlàg $\mathbb{F}$-martingale with terminal value $\mathbb{P}(C_n|\mathscr{F}_\infty)$, namely

$$z_t^n := \mathbb{P}(C_n|\mathscr{F}_t). \tag{5.10}$$

The following result describes how, for a thin time, conditional expectations given $\mathscr{G}_t$ can be expressed in terms of conditional expectations given $\mathscr{F}_t$.

Lemma 5.32 *Let τ be a thin time with exhausting sequence $(T_n)_{n\geq 1}$ and $(z^n)_{n\geq 1}$ be the family of $\mathbb{F}$-martingales associated with τ through* (5.10). *Then, for any $\mathscr{F}$-measurable integrable random variable X and $s \leq t$, we have*

$$\mathbb{E}\left[X|\mathscr{G}_t\right] \mathbb{1}_{\{s\geq T_n\}\cap C_n} = \mathbb{1}_{\{s\geq T_n\}\cap C_n} \frac{\mathbb{E}\left[X \mathbb{1}_{C_n}|\mathscr{F}_t\right]}{z_t^n}.$$

Proof Note that

$$\mathscr{G}_t = \bigcap_{u>t} \mathscr{F}_u \vee \sigma(C_n \cap \{T_n \leq s\},\ s \leq u,\ n \in \mathbb{N}).$$

Thus, by the monotone class theorem, for each $G \in \mathscr{G}_t$ there exists $F \in \mathscr{F}_t$ such that

$$G \cap \{T_n \leq s\} \cap C_n = F \cap \{T_n \leq s\} \cap C_n. \tag{5.11}$$

Then, we have to show that

$$\mathbb{E}\left[X\, z_t^n \mathbb{1}_{\{s\geq T_n\}\cap C_n}|\mathscr{G}_t\right] = \mathbb{1}_{\{s\geq T_n\}\cap C_n}\mathbb{E}\left[X \mathbb{1}_{\{s\geq T_n\}\cap C_n}|\mathscr{F}_t\right].$$

For any $G \in \mathscr{G}_t$, we choose $F \in \mathscr{F}_t$ satisfying (5.11), and we obtain

$$\begin{aligned}
&\mathbb{E}\left[X\, z_t^n \mathbb{1}_{\{s\geq T_n\}\cap C_n \cap G}\right] = \mathbb{E}\left[X \mathbb{1}_{\{s\geq T_n\}\cap C_n\cap F}\ \mathbb{E}\left[\mathbb{1}_{C_n}|\mathscr{F}_t\right]\right]\\
&= \mathbb{E}\left[\mathbb{1}_{\{s\geq T_n\}\cap F}\ \mathbb{E}\left[\mathbb{1}_{C_n}|\mathscr{F}_t\right]\mathbb{E}\left[X\mathbb{1}_{C_n}|\mathscr{F}_t\right]\right] = \mathbb{E}\left[\mathbb{1}_{\{s\geq T_n\}\cap C_n\cap F}\ \mathbb{E}\left[X\mathbb{1}_{C_n}|\mathscr{F}_t\right]\right]\\
&= \mathbb{E}\left[\mathbb{1}_{\{s\geq T_n\}\cap C_n\cap G}\ \mathbb{E}\left[X\mathbb{1}_{C_n}|\mathscr{F}_t\right]\right]
\end{aligned}$$

which ends the proof. □

Let $\mathbb{F}^{\mathscr{C}}$ be the initial enlargement of the filtration $\mathbb{F}$ with the atomic σ-field $\mathscr{C} := \sigma(C_n, n \geq 0)$ with C_n defined in (5.9), i.e.,

$$\mathscr{F}_t^{\mathscr{C}} := \bigcap_{s>t} \mathscr{F}_s \vee \sigma(C_n, n \geq 0). \tag{5.12}$$

Equivalently, if $\zeta := \sum_{n=0}^{\infty} n \mathbb{1}_{C_n}$, then $\mathbb{F}^{\mathscr{C}} = \mathbb{F}^{\sigma(\zeta)}$. Thus, by Theorem 4.25 and Example 4.15, hypothesis $\mathscr{H}'$ between $\mathbb{F}$ and $\mathbb{F}^{\mathscr{C}}$ holds and the decomposition of any $\mathbb{F}$-local martingale X as an $\mathbb{F}^{\mathscr{C}}$-semimartingale is

$$X_t = \widehat{X}_t + \sum_n \mathbb{1}_{C_n} \int_0^t \frac{1}{z_{s-}^n} d\langle X, z^n\rangle_s^{\mathbb{F}}, \tag{5.13}$$

where $\widehat{X}$ is an $\mathbb{F}^{\mathscr{C}}$-local martingale and z^n are given in (5.10).

Theorem 5.33 *Let τ be a thin time. Then $\mathbb{F} \subset \mathbb{G} \subset \mathbb{F}^{\mathscr{C}}$ and the hypothesis $\mathscr{H}'$ between $\mathbb{F}$ and $\mathbb{G}$ is satisfied. Moreover, any $\mathbb{F}$-local martingale X has the fol-*

lowing $\mathbb{G}$-semimartingale canonical decomposition

$$X_t = \widehat{X}_t + \int_0^{t\wedge\tau} \frac{1}{Z_{s-}} d\langle X, m\rangle_s^{\mathbb{F}} + \sum_n \mathbb{1}_{C_n} \int_0^t \mathbb{1}_{\{s>T_n\}} \frac{1}{z_{s-}^n} d\langle X, z^n\rangle_s^{\mathbb{F}} \tag{5.14}$$

where $\widehat{X}$ is a $\mathbb{G}$-local martingale.

Proof It is straightforward that $\mathbb{F} \subset \mathbb{G} \subset \mathbb{F}^{\mathscr{C}}$. Thus it follows by (5.13) and Stricker's Theorem 1.25 that the hypothesis $\mathscr{H}'$ between $\mathbb{F}$ and $\mathbb{G}$ is satisfied. To derive the decomposition, let H be a $\mathbb{G}$-predictable bounded process. Then, Proposition 2.11 (b) implies that

$$H = \mathbb{1}_{[\![0,\tau]\!]} J + \mathbb{1}_{]\!]\tau,\infty[\![} K(\tau), \quad t \geq 0$$

where J is the $(\mathbb{F}, \tau)$-predictable reduction of H and $K : \Omega \times \mathbb{R}^+ \times \overline{\mathbb{R}}^+ \to \mathbb{R}$ is a bounded $\mathscr{P}\otimes\mathscr{B}(\overline{\mathbb{R}}^+)$-measurable mapping. In particular, J is bounded $\mathbb{F}$-predictable process satisfying $J = J\mathbb{1}_{\{Z_->0\}}$. Since τ is a thin time, we can rewrite the process H as

$$H = J\mathbb{1}_{[\![0,\tau]\!]} + \sum_n \mathbb{1}_{]\!]T_n,\infty[\![} K(T_n)\mathbb{1}_{C_n}$$

with $C_n = \{\tau = T_n < \infty\}$. Note that each process $K^n := \mathbb{1}_{]\!]T_n,\infty[\![} K(T_n)$ is $\mathbb{F}$-predictable and bounded and, since $C_n \subset \{z_{t-}^n > 0\}$, K^n can be chosen to satisfy $K_t^n = K_t^n \mathbb{1}_{\{z_{t-}^n>0\}}$.

Let X be an $H^1(\mathbb{F})$-martingale. Then the stochastic integrals $J \cdot X$ and $K^n \cdot X$ are well-defined and each of them is an $H^1(\mathbb{F})$-martingale. For each $n \in \mathbb{N}$ and for each bounded $\mathbb{F}$-martingale N, by integration by parts, we have that

$$\mathbb{E}\left[\mathbb{1}_{C_n} N_\infty\right] = \mathbb{E}\left[[z^n, N]_\infty\right] = \mathbb{E}\left[\langle z^n, N\rangle_\infty^{\mathbb{F}}\right]. \tag{5.15}$$

Since $N \to \mathbb{E}[\mathbb{1}_{C_n} N_\infty]$ is a linear form, the duality (H^1, BMO) implies that (5.15) holds for any $H^1(\mathbb{F})$-martingale N. Similarly, by Proposition 1.49 (b), for any $H^1(\mathbb{F})$-martingale N, the process $\langle N, m\rangle^{\mathbb{F}}$ exists and we have

$$\mathbb{E}[N_\tau] = \mathbb{E}\left[[N, m]_\infty\right] = \mathbb{E}\left[\langle N, m\rangle_\infty^{\mathbb{F}}\right]$$

where m is given in Proposition 1.46 (b). Therefore

$$\begin{aligned}\mathbb{E}\left[\int_0^\infty H_s dX_s\right] &= \mathbb{E}\left[\int_0^\tau J_s dX_s\right] + \sum_n \mathbb{E}\left[\mathbb{1}_{C_n} \int_0^\infty K_s^n dX_s\right] \\ &= \mathbb{E}\left[\int_0^\infty J_s d\langle m, X\rangle_s^{\mathbb{F}}\right] + \sum_n \mathbb{E}\left[\int_0^\infty K_s^n d\langle z^n, X\rangle_s^{\mathbb{F}}\right].\end{aligned}$$

Then, from (1.9) we deduce

$$
\begin{aligned}
\mathbb{E}\left[\int_0^\infty H_s dX_s\right] &= \mathbb{E}\left[\int_0^\infty \frac{Z_{s-}}{Z_{s-}} 1\!\!1_{\{Z_{s-}>0\}} J_s d\langle m, X\rangle_s^{\mathbb{F}}\right] \\
&\quad + \sum_n \mathbb{E}\left[\int_0^\infty \frac{z^n_{s-}}{z^n_{s-}} 1\!\!1_{\{z^n_{s-}>0\}} K^n_s d\langle z^n, X\rangle_s^{\mathbb{F}}\right] \\
&= \mathbb{E}\left[\int_0^\tau \frac{1}{Z_{s-}} J_s d\langle m, X\rangle_s^{\mathbb{F}}\right] + \sum_n \mathbb{E}\left[1\!\!1_{C_n} \int_0^\infty \frac{1}{z^n_{s-}} K^n_s d\langle z^n, X\rangle_s^{\mathbb{F}}\right].
\end{aligned}
$$

For any $H^1(\mathbb{F})$-martingale X, the assertion of the theorem follows as, for any $s \le t$ and $F \in \mathscr{G}_s$, the process $H = 1\!\!1_{(s,t]} 1\!\!1_F$ is $\mathbb{G}$-predictable. To conclude the proof, we recall that any local martingale is locally in H^1. □

5.5 Pseudo-Stopping Times

It is well known, and described in Sect. 3.2.1, that the property that $\mathbb{F} \hookrightarrow \mathbb{G}$ implies that the Azéma supermartingales Z and $\widetilde{Z}$ associated with τ are decreasing processes. However, the converse implication does not hold. It turns out that monotonicity of Z and $\widetilde{Z}$ is closely related to the property that τ is a pseudo-stopping time.

Definition 5.34 *A random time τ is an $\mathbb{F}$-pseudo-stopping time if, for any bounded $\mathbb{F}$-martingale Y, $\mathbb{E}[Y_\tau] = \mathbb{E}[Y_0]$.*

In particular, any $\mathbb{F}$-stopping time is a pseudo-stopping time, by the optional sampling theorem, since we only consider bounded martingales in Definition 5.34.

The following result collects equivalent characterisations of a pseudo-stopping time property.

Theorem 5.35 *Let τ be a random time with associated processes defined in Proposition 1.46. Then the following conditions are equivalent:*

(a) τ is an $\mathbb{F}$-pseudo-stopping time;
(b) $A^o_\infty = \mathbb{P}(\tau < \infty | \mathscr{F}_\infty)$;
(c) $m = 1$;
(d) ${}^oA = A^o$;
(e) for every $\mathbb{F}$-local martingale Y, the process Y^τ is a $\mathbb{G}$-local martingale.
(f) $\widetilde{Z}$ is a decreasing càglàd process.

Proof Let Y be a bounded $\mathbb{F}$-martingale. Then, Proposition 1.49 implies that $\mathbb{E}[Y_\tau] = \mathbb{E}\,[[Y, m]_\infty]$ which shows that (c) implies (a) as $m_0 = 1$. By applying Proposition 1.49 to m, (a) implies that $\mathbb{E}\,[[m, m]_\infty] = m_0 = 1$ from which we conclude that m is a constant martingale, thus (c).

The equivalence between (c) and (d) follows by Proposition 1.46 (b).

The implication (e) ⇒ (a) is straightforward, while the implication (c) ⇒ (e) comes from general decomposition result for stopped martingales stated in (5.1). The equivalence of (c) and (f) comes from $\widetilde{Z} = m - A^o_-$. Condition (f) implies that m is a continuous (as it is càdlàg and càglàd) finite variation martingale, thus it is a constant, $m = 1$. □

We further investigate the relation between pseudo-stopping times and immersion for a general enlargement of filtration in the following theorem.

Theorem 5.36 *Given two filtrations $\mathbb{F}$ and $\mathbb{H}$ such that $\mathbb{F} \subset \mathbb{H}$, the following conditions are equivalent.*

(a) The filtration $\mathbb{F}$ is immersed in $\mathbb{H}$.
(b) Every $\mathbb{H}$-stopping time is an $\mathbb{F}$-pseudo-stopping time.
(c) The $\mathbb{F}$-dual optional projection of any $\mathbb{H}$-optional process of integrable variation is equal to its $\mathbb{F}$-optional projection.

Proof We start with showing (a) $\Rightarrow$ (b). Let Y be a bounded $\mathbb{F}$-martingale and ϑ an $\mathbb{H}$-stopping time. Then, (a) implies that Y is a bounded $\mathbb{H}$-martingale, therefore $\mathbb{E}[Y_\vartheta] = \mathbb{E}[Y_0]$, which implies that ϑ is an $\mathbb{F}$-pseudo-stopping time.

To show (b) $\Rightarrow$ (a), suppose that Y is a bounded $\mathbb{F}$-martingale and ϑ is an $\mathbb{H}$-stopping time. Since, by hypothesis, every $\mathbb{H}$-stopping time is an $\mathbb{F}$-pseudo-stopping time, we have $\mathbb{E}[Y_\vartheta] = \mathbb{E}[Y_0]$ for every $\mathbb{G}$-stopping time ϑ, which by Theorem 1.42 in [125], implies that Y is an $\mathbb{H}$-martingale.

The implication (c) $\Rightarrow$ (b) follows directly from Theorem 5.35 (d). For the proof of the implication (a) $\Rightarrow$ (c) we refer to [12]. □

Example 5.37 The first example of a pseudo-stopping time was given by Williams [210]. Let B be a Brownian motion with its natural filtration $\mathbb{F}$ and define the $\mathbb{F}$-stopping time $T_1 = \inf\{t : B_t = 1\}$ and the random time $\vartheta = \sup\{t < T_1 : B_t = 0\}$ (see Example 5.7 (a)). Set

$$\tau = \sup\{s < \vartheta : B_s = B_s^*\}$$

where B^* is the running maximum of the Brownian motion. Then, τ is proven to be a pseudo-stopping time. Note that $\mathbb{E}[B_\tau]$ is not equal to 0; this illustrates the fact we cannot take any martingale in Definition 5.34. The martingale $(B_{t\wedge T_1}, t \geq 0)$ is neither bounded, nor uniformly integrable. In fact, since the maximum B_ϑ^* $(= B_\tau)$ is uniformly distributed on [0, 1], one has $\mathbb{E}[B_\tau] = 1/2$.

Example 5.38 We present another example where τ is a pseudo stopping-time. Let S be defined through $dS_t = \sigma S_t dB_t$, where B is a Brownian motion and σ a constant. Let $\tau = \sup\{t \leq 1 : S_1 - 2S_t = 0\}$, that is the last time before 1 at which the process S is equal to half of its terminal value at time 1. Note that

$$\{\tau \leq t\} = \{\inf_{t\leq s\leq 1} 2S_s \geq S_1\} = \left\{\inf_{t\leq s\leq 1} 2\frac{S_s}{S_t} \geq \frac{S_1}{S_t}\right\}.$$

Since $\left(\frac{S_s}{S_t}, s \geq t\right)$ and $\frac{S_1}{S_t}$ are independent of $\mathscr{F}_t$, and using the fact that $\left(\frac{S_s}{S_t}, s \geq t\right)$ and $(S_{s-t}, s \geq t)$ have the same law, we obtain

$$\mathbb{P}\left(\inf_{t\le s\le 1} 2\frac{S_s}{S_t} \ge \frac{S_1}{S_t}|\mathscr{F}_t\right) = \mathbb{P}\left(\inf_{t\le s\le 1} 2S_{s-t} \ge S_{1-t}\right) = \Phi(1-t)$$

where $\Phi(u) = \mathbb{P}(\inf_{s\le u} 2S_s \ge S_u)$. It follows that the supermartingale $\widetilde{Z} = Z$ is a deterministic continuous decreasing function, hence τ is a pseudo-stopping time. In this example, one can also show that any $\mathbb{F}$-martingale X is a $\mathbb{G}$-semimartingale (see [4, pp. 83–84]).

Finally, we relate pseudo-stopping times with honest times.

Proposition 5.39 *For a random time τ, the following conditions are equivalent*
(a) τ is equal to an $\mathbb{F}$-stopping time on $\{\tau < \infty\}$.
(b) τ is an $\mathbb{F}$-pseudo-stopping time and an $\mathbb{F}$-honest time.
In particular if τ is an $\mathbb{F}$-honest time which is not equal to an $\mathbb{F}$-stopping time on $\{\tau < \infty\}$, then $\mathbb{F}$ is not immersed in $\mathbb{G}$.

Proof The implication (a) $\Rightarrow$ (b) is trivial. Assume (b) holds. Then, by Theorem 5.8 (c), $\tau = \sup\{t : \widetilde{Z}_t = 1\}$ on $\{\tau < \infty\}$ and, by Theorem 5.35 (d), $Z = 1 - A^o$. Moreover we use the general relation $\widetilde{Z} - Z = \Delta A^o$. Then, on $\{\tau < \infty\}$ we obtain

$$\begin{aligned}\tau = \sup\{t : \widetilde{Z}_t = 1\} &= \sup\{t : Z_t + \Delta A^o_t = 1\} \\ &= \sup\{t : 1 - A^o_t + \Delta A^o_t = 1\} = \sup\{t : A^o_{t-} = 0\} = \inf\{t : A^o_t > 0\}.\end{aligned}$$

So, τ equals an $\mathbb{F}$-stopping time on $\{\tau < \infty\}$. Therefore if τ is an $\mathbb{F}$-honest time which is not equal to an $\mathbb{F}$-stopping time on $\{\tau < \infty\}$ then, by Theorem 5.36, $\mathbb{F}$ is not immersed in $\mathbb{G}$. □

5.6 Predictable Representation Property

The predictable representation property in the enlarged filtration $\mathbb{G}$ is not guaranteed, even if it holds for the reference filtration $\mathbb{F}$. We recall here, without proofs, some of the results found in the literature concerning this problem. We assume that $\mathbb{F}$ enjoys the predictable representation property w.r.t. a finite family of $\mathbb{F}$-local martingales $Y = (Y^1, \ldots, Y^n)$, and we denote by $\widehat{Y} = (\widehat{Y}^1, \ldots, \widehat{Y}^n)$ the $\mathbb{G}$-local martingale parts in the canonical $\mathbb{G}$-semimartingale decomposition of Y (assumed to exist). As before in (2.10), by M we denote the $\mathbb{G}$-compensated martingale associated with the process A, namely $M = A - (1 - A_-)\frac{1}{Z_-} \cdot A^p$. We recall that the multiplicity of a filtration $\mathbb{F}$ is, if it exists, the smallest number $n \in \mathbb{N}$ such that there exists a family of n one-dimensional $\mathbb{F}$-local martingales $(Y^1, \ldots, Y^n)$ such that $(Y^1, \ldots, Y^n)$ has the PRP in $\mathbb{F}$.

1. The following two conditions are equivalent [134].
   (a) The filtration $\mathbb{F}$ is immersed in $\mathbb{G}$ and Y_τ is $\mathscr{F}_{\tau-}$-measurable.
   (b) $(\widehat{Y}, M)$ has the PRP in $\mathbb{G}$ and $\mathscr{G}_\tau = \mathscr{G}_{\tau-}$.

2. Under Jacod's absolute continuity condition, $(\widehat{Y}, M)$ has the PRP in $\mathbb{G}$ [99].
3. If $\mathbb{F}$ satisfies condition **(C)** and τ is honest, then all continuous $\mathbb{G}$-martingales are generated by the family $\widehat{Y}$ [28].
4. If τ is an honest time and condition **(C)** holds, then, for any bounded $\mathbb{G}$-martingale X, there exist $\mathbb{F}$-predictable processes J', J'', K and a bounded $\mathcal{G}_\tau$-measurable r.v. ξ satisfying $\mathbb{E}[\xi|\mathcal{G}_{\tau-}] = 0$, such that

$$X = X_0 + \left(J' 1\!\!1_{[\![0,\tau]\!]} + J'' 1\!\!1_{]\!]\tau,\infty[\![}\right) \cdot \widehat{Y} + K 1\!\!1_{]\!]0,\tau]\!]} \cdot M + 1\!\!1_{\{\tau>0\}} \xi A.$$

   If, in addition, $\{0 < \tau < \infty\} \cap \mathcal{G}_{\tau-} = \{0 < \tau < \infty\} \cap \mathcal{G}_\tau$, the predictable representation property holds in $\mathbb{G}$ w.r.t. $(\widehat{Y}, M)$. The multiplicity of $\mathbb{F}$ is n, and the multiplicity of $\mathbb{G}$ is $n+1$ [139, Theorem 5.12] and [134].
5. If τ is an honest time and $\mathbb{F}$ is a Brownian filtration, there exists a bounded $\mathbb{G}$-martingale X such that $(\widehat{Y}, M, X)$ has the predictable representation property in $\mathbb{G}$ (the multiplicity of $\mathbb{F}$ is n, and the multiplicity of $\mathbb{G}$ is $n+2$ - or $n+1$ if $X = 0$) [134].
6. If τ is an honest time, the space of square integrable $\mathbb{G}$-martingales is generated by two families [28]:
   - the family $\left\{\widehat{X} : \widehat{X} \text{ is the } \mathbb{G}\text{-martingale part of a bounded } \mathbb{F}\text{-martingale } X\right\}$,
   - and the family $\left\{v 1\!\!1_{[\![\tau,\infty[\![} - (v 1\!\!1_{[\![\tau,\infty[\![})^{\mathbb{G},p} : v \text{ is bounded } \mathcal{G}_\tau\text{-measurable r.v.}\right\}$.

5.7 Enlargement with a Future Infimum Process

In this section we examine the progressive enlargement of a filtration $\mathbb{F}$ with a future infimum (or supremum) process. This is a qualitatively different example than all constructions presented so far in this book.

Proposition 5.40 *Let Y be a non-negative continuous $\mathbb{F}$-local martingale such that $Y_0 > 0$ and $Y_t \to 0$ when $t \to \infty$ a.s. Let $\Sigma_t := \sup_{u \geq t} Y_u$ and $\widetilde{\mathbb{F}} := \mathbb{F} \vee \mathbb{F}^\Sigma$ where $\mathbb{F}^\Sigma$ is the natural filtration of Σ. Then,*

$$Y = Y_0 + 2\Sigma + \frac{1}{Y} \cdot \langle Y \rangle^{\mathbb{F}} + \widetilde{Y}$$

where $\widetilde{Y}$ is an $\widetilde{\mathbb{F}}$-local martingale.

Proof Let $T_0 := \inf\{t : Y_t = 0\}$ and define $C := Y^{-4} \cdot \langle Y \rangle^{\mathbb{F}}$. Then Itô's formula and the Dambis–Dubins–Schwarz theorem (see [190, p. 181]) imply that

$$\frac{1}{Y_t} = R_{C_t}, \quad t < T_0 \tag{5.16}$$

where R is a 3-dimensional Bessel process, in the filtration $\mathbb{K} := (\mathscr{K}_t, t \geq 0)$, where $\mathscr{K}_t = \mathscr{F}_{\theta_t}$ and θ is the right-inverse of C. We now consider the enlarged filtration $\widetilde{\mathbb{K}} := (\widetilde{\mathscr{K}_t}, t \geq 0)$ where $\widetilde{\mathscr{K}_t} = \mathscr{K}_t \vee \sigma(J_t)$ and $J_t = \inf_{s \geq t} R_s$ (it is easy to check that, indeed $\widetilde{\mathbb{K}}$ is a filtration). The equality (5.16) implies $\Sigma_t^{-1} = J_{C_t}$ and Pitman's theorem [190, Theorem (3.5) Chap. VI, p.253] yields

$$\frac{1}{Y_t} = \frac{1}{Y_0} + 2\Sigma_t^{-1} - \widetilde{X}_t$$

where $\widetilde{X}$ is an $\widetilde{\mathbb{F}}$-local martingale. From Itô's formula, working in $\widetilde{\mathbb{F}}$, one has

$$\begin{aligned} Y_t &= Y_0 - \int_0^t Y_s^2 d\frac{1}{Y_s} + \int_0^t Y_s^3 d\langle \frac{1}{Y} \rangle_s = Y_0 - \int_0^t Y_s^2 d\left(\frac{2}{\Sigma_s} - \widetilde{X}_s\right) + \int_0^t Y_s^3 \frac{d\langle Y \rangle_s}{Y_s^4} \\ &= Y_0 + 2\int_0^t \frac{Y_s^2}{\Sigma_s^2} d\Sigma_s + \int_0^t \frac{d\langle Y \rangle_s}{Y_s} + \widetilde{Y}_t \end{aligned}$$

where $\widetilde{Y}_t = \int_0^t Y_s^2 d\widetilde{B}_s$ is an $\widetilde{\mathbb{F}}$-local martingale. □

Remark 5.41 More generally, it is possible to enlarge the filtration of a diffusion satisfying $Y_t \to \infty$ as $t \to \infty$ and

$$Y_t = y + B_t + \int_0^t c(Y_u) du,$$

where B is a Brownian motion, with Y's future infimum, i.e., to consider Y in the filtration $\widetilde{\mathscr{Y}_t} := \mathscr{Y}_t \vee \sigma(J_t)$ where $J_t = \inf_{s \geq t} Y_s$. Note that $(\widetilde{\mathscr{Y}_t}, t \geq 0)$ is indeed a filtration. There exists an $(\widetilde{\mathscr{Y}_t}, t \geq 0)$ Brownian motion β such that

$$Y_t = y + \beta_t + \int_0^t \left(\frac{s'}{s} - \frac{1}{2}\frac{s''}{s'}\right)(Y_u) du + 2J_t$$

where s is a scale function of Y that satisfies $s(0) = -\infty$, $s(\infty) = 0$, $\frac{1}{2}s'' + cs' = 0$.

5.8 Arbitrages in a Progressive Enlargement

This section contains a study of the existence of arbitrages arising due to the enlargement of filtration. We do not give the broadest results, due to the length of the proofs in a general setting. We refer the reader to *Bibliographic Notes* given in the last section of this chapter for more information.

5.8.1 Classical Arbitrages for Honest Times

We consider a finite honest time and a financial market, consisting of a savings account with null interest rate and a risky asset with price process S. We make use of the standard definitions on classical arbitrages recalled in Sect. 1.4.

Theorem 5.42 *Let $[0, T\rangle$ be the time horizon. Assume that $(S, \mathbb{F})$ is a complete market satisfying NFLVR. If τ is a positive $[0, T\rangle$-valued honest time which is not an $\mathbb{F}$-stopping time, then there are classical arbitrages for $(S^\tau, \mathbb{G})$ and $(S - S^\tau, \mathbb{G})$.*

Proof We assume w.l.o.g. that S is an $(\mathbb{F}, \mathbb{P})$-martingale since we can work under the EMM. We provide two different constructions of arbitrage opportunities.

A first construction is given under **(CA)** conditions. In that case, from Theorem 5.16, one has $Z = N/N^*$, where N is a non-negative $\mathbb{F}$-local martingale with $N_0 = 1$ and N^* is non-decreasing with $N_0^* = 1$, hence $N_t^* \geq 1$ for any t. The market being complete, there exists a strategy φ such that $N_t = 1 + \int_0^t \varphi_s dS_s$. Since $N \geq 0$, the strategy is admissible and $N - 1$ is the wealth associated with φ and null initial value. Theorem 5.8 (c) and the fact that $Z = \widetilde{Z}$ imply that $\tau = \sup\{t : N_t = N_t^*\}$, hence $N_\tau = N_\tau^* = N_T^*$, $N_\tau - 1 \geq 0$ and $\mathbb{P}(N_\tau - 1 > 0) > 0$. It is now clear that using the strategy φ up to time τ provides an arbitrage. One has also, for $t > \tau$, $N_\tau - N_t = N_T^* - N_t \geq 0$. Since $N_\tau - N_t = \int_\tau^t (-\varphi_s) dS_s$, the strategy $-\varphi$ is admissible and is a $\mathbb{G}$ arbitrage after τ.

The second construction is given for the general case. We use the fact that for an honest time $A^o_\cdot = A^o_{\cdot\wedge\tau}$ and $\widetilde{Z}_\tau = 1$. From $m = \widetilde{Z} + A^o_-$, we deduce that $m_\tau \geq 1$. Due to the completeness of the market, there exists ψ such that $m_t - 1 = \int_0^t \psi_s dS_s$ and $m - 1$ is a wealth process with null initial value. Since τ is not an $\mathbb{F}$-stopping time, one has $\mathbb{P}(A^o_{\tau-} > 0) > 0$ and $\mathbb{P}(m_\tau > 1) > 0$, hence ψ leads to an arbitrage at time τ. Using $m = \widetilde{Z} + A^o_-$, one obtains that, for $t > \tau$, $m_t - m_\tau = \widetilde{Z}_t - 1 + \Delta A^o_\tau \geq -1$. Consider the following $\mathbb{G}$-stopping time, which is $[0, T\rangle$-valued,

$$\vartheta := \inf\left\{t > \tau : \widetilde{Z}_t \leq \frac{1 - \Delta A^o_\tau}{2}\right\}. \tag{5.17}$$

Then,

$$m_\vartheta - m_\tau = \widetilde{Z}_\vartheta - 1 + \Delta A^o_\tau \leq \frac{\Delta A^o_\tau - 1}{2} \leq 0,$$

and, as τ is not an $\mathbb{F}$-stopping time,

$$\mathbb{P}(m_\vartheta - m_\tau < 0) = \mathbb{P}(\Delta A^o_\tau < 1) > 0.$$

Hence $-\int_\tau^{t\wedge\vartheta} \psi_s dS_s = m_{\tau\wedge t} - m_{t\wedge\vartheta}$ is the value of an admissible strategy with initial value 0 and terminal value $m_\tau - m_\vartheta \geq 0$ satisfying $\mathbb{P}(m_\tau - m_\vartheta > 0) > 0$. This ends the proof of the theorem. □

• **Examples in a Brownian Filtration**

Let $[0, \infty)$ be the time horizon and let $dS_t = S_t \sigma dB_t,\ S_0 > 0$ be the risk neutral price of a risky asset in a financial market with null interest rate.
(a) **Last passage time**. Consider, for $a \in (0, S_0)$, the finite $\mathbb{F}$-honest time

$$\tau := \sup\{t \geq 0 : S_t = a\},$$

which satisfies condition **(A)**. From Proposition 5.19, the associated Azéma supermartingale is given by $Z_t = (S_t/a) \wedge 1$. By the Itô–Tanaka formula, the Doob–Meyer decomposition of Z is

$$Z_t = m_t - A^o_t = 1 + \frac{1}{a}\int_0^t \mathbb{1}_{\{S_u < a\}} dS_u - \frac{1}{2a} L^a_t,$$

where $L^a = (L^a_t)_{t \geq 0}$ denotes the local time of S at the level a. It follows from the second construction above that the $\mathbb{G}$-predictable strategy $\mathbb{1}_{[\![0,\tau]\!]}\frac{1}{a}\mathbb{1}_{\{S<a\}}$ yields an arbitrage opportunity.

The martingale part of the multiplicative decomposition of Z is given by

$$N_t = \mathscr{E}\left(\frac{1}{aZ}\mathbb{1}_{\{S<a\}} \cdot S\right)_t = 1 + \frac{1}{a}\int_0^t e^{L^a_s/(2a)} \mathbb{1}_{\{S_s < a\}} dS_s.$$

The $\mathbb{G}$-predictable strategy $\varphi := \mathbb{1}_{[\![0,\tau]\!]}(e^{L^a/(2a)}/a)\mathbb{1}_{\{S<a\}}$ yields an arbitrage opportunity in the enlarged financial market. After τ, the strategy $-\varphi$ is an arbitrage.
(b) **Last passage time before maturity**. Consider the following honest time

$$\tau := \sup\{t \leq 1 \ :\ S_t = a\}$$

where $a \in (0, S_0)$. Setting $V_t := \alpha - \gamma t - B_t$, with $\alpha = \frac{\ln a}{\sigma}$ and $\gamma = -\frac{\sigma}{2}$, one has

$$\tau = \sup\{t \leq 1 \ :\ \gamma t + B_t = \alpha\} = \sup\{t \leq 1 \ :\ V_t = 0\}.$$

Setting $T_0(V) = \inf\{t \ :\ V_t = 0\}$, we obtain, using standard computations (see [136, p. 145–148])

$$1 - Z_t = \left(1 - e^{\gamma V_t} H(\gamma, |V_t|, 1-t)\right) \mathbb{1}_{\{T_0(V) \leq t \leq 1\}} + \mathbb{1}_{\{t > 1\}},$$

where

$$H(z, y, s) := e^{-zy}\mathscr{N}\left(\frac{zs - y}{\sqrt{s}}\right) + e^{zy}\mathscr{N}\left(\frac{-zs - y}{\sqrt{s}}\right),$$

where $\mathscr{N}(x)$ is the cumulative distribution function of the standard normal distribution. Using Itô's lemma, we obtain the decomposition of $1 - e^{\gamma V_t} H(\gamma, |V_t|, 1-t)$ as a semimartingale. The martingale part of Z is of the form $dm_t = \varphi_t dS_t$ where φ is given explicitly in [5].

• Examples in a Poisson Filtration

Let $[0,\infty)$ be the time horizon. We suppose that N is a Poisson process, with intensity $\lambda > 0$, and $\mathbb{F}$ is its natural filtration. The stock price process is given by

$$dS_t = S_{t-}\sigma(dN_t - \lambda dt), \quad S_0 = 1,$$

with $\sigma > -1$, or equivalently $S_t = \exp(-\lambda\sigma t + \ln(1+\sigma)N_t)$. For $b \in (0,1)$ we introduce the following notation

$$\alpha := \ln(1+\sigma) > 0, \quad a := -\frac{1}{\alpha}\ln b, \quad \mu := \frac{\lambda\sigma}{\ln(1+\sigma)} \text{ and } Y_t := \mu t - N_t .$$

Thus $S_t = \exp(-\ln(1+\sigma)Y_t)$. To the process Y, we associate its ruin probability, denoted by $\psi(x)$ and given by

$$\psi(x) = \mathbb{P}(T^x < \infty), \quad \text{with} \quad T^x = \inf\{t : x + Y_t < 0\} \quad \text{and} \quad x \geq 0. \quad (5.18)$$

It is known [22, Chap. III, cor. 3.1] that $\psi(0) = (1+\theta)^{-1}$ where $\theta = \dfrac{\mu}{\lambda} - 1$.

Proposition 5.43 *For $0 < b < 1$, consider the following random time*

$$\tau := \sup\{t : S_t \geq b\} = \sup\{t : Y_t \leq a\}.$$

Suppose that $\theta > 0$. Then the following assertions hold:
(a) τ is a finite thin honest time.
(b) The process

$$\varphi := \frac{1}{\sigma S_-}\left(\psi(Y_- - a - 1)\mathbb{1}_{\{Y_- \geq a+1\}} - \psi(Y_- - a)\mathbb{1}_{\{Y_- \geq a\}} + \mathbb{1}_{\{Y_- < a+1\}} - \mathbb{1}_{\{Y_- < a\}}\right)$$

is an arbitrage opportunity for $(S^\tau, \mathbb{G})$, and $-\varphi I_{]\!]\tau,\vartheta]\!]}$ is an arbitrage opportunity for $(S - S^\tau, \mathbb{G})$. Here, the function ψ is defined in (5.18) and ϑ is defined in the same manner as in (5.17).

Proof Since $\theta > 0$, one has $\mu > \lambda$ so $Y_t \to \infty$ as $t \to \infty$, and τ is finite. The Azéma supermartingale associated with the time τ is

$$Z_t = \mathbb{P}(\tau > t|\mathcal{F}_t) = \psi(Y_t - a)\mathbb{1}_{\{Y_t \geq a\}} + \mathbb{1}_{\{Y_t < a\}} = 1 + \mathbb{1}_{\{Y_t \geq a\}}\left(\psi(Y_t - a) - 1\right),$$

where ψ is defined in (5.18). Since $Y_\tau = a$, we obtain $Z_\tau = \frac{1}{1+\theta} < 1$.

Define $\vartheta_1 := \inf\{t > 0 : Y_t = a\}$ and, for any $n > 1$, $\vartheta_n = \inf\{t > \vartheta_{n-1} : Y_t = a\}$. It can be proven that the times ϑ_n are $\mathbb{F}$-predictable stopping times, and $[\![\tau]\!] \subset \cup_n [\![\vartheta_n]\!]$. Hence, the random time τ is thin. For any optional increasing process K, one has

$$\mathbb{E}[K_\tau] = \mathbb{E}\left[\sum_{n=1}^{\infty} \mathbb{1}_{\{\tau=\vartheta_n\}} K_{\vartheta_n}\right] = \mathbb{E}\left[\sum_{n=1}^{\infty} \mathbb{E}\left[\mathbb{1}_{\{\tau=\vartheta_n\}}|\mathscr{F}_{\vartheta_n}\right] K_{\vartheta_n}\right]$$

and $\mathbb{E}[\mathbb{1}_{\{\tau=\vartheta_n\}}|\mathscr{F}_{\vartheta_n}] = \mathbb{P}(T^0 = \infty) = 1 - \Psi(0) = \frac{\theta}{1+\theta}$. It follows that the $\mathbb{F}$-dual optional projection A^o of the process $\mathbb{1}_{[\![\tau,\infty[\![}$ is

$$A^o = \frac{\theta}{1+\theta}\sum_n \mathbb{1}_{[\![\vartheta_n,\infty[\![}.$$

Note that $\widetilde{Z}_\tau = Z_\tau + \Delta A^o_\tau = 1 + (\psi(0) - 1) + \frac{\theta}{1+\theta} = 1$, hence τ is honest.

As a result, the process A^o is $\mathbb{F}$-predictable, and hence we have that $Z = m - A^o$ is the Doob–Meyer decomposition of Z. Thus, we deduce

$$\Delta m = Z - {}^pZ$$

where ${}^pZ$ is the predictable projection of Z. To calculate ${}^pZ$, we write the process Z in a more adequate form. To this end, we first remark that

$$\begin{aligned}\mathbb{1}_{\{Y\geq a\}} &= \mathbb{1}_{\{Y_-\geq a+1\}}\Delta N + (1 - \Delta N)\mathbb{1}_{\{Y_-\geq a\}}\\ \mathbb{1}_{\{Y<a\}} &= \mathbb{1}_{\{Y_-<a+1\}}\Delta N + (1 - \Delta N)\mathbb{1}_{\{Y_-<a\}}.\end{aligned}$$

Then, we obtain easily

$$\begin{aligned}\Delta m &= \left(\psi(Y_- - a - 1)\mathbb{1}_{\{Y_-\geq a+1\}} - \psi(Y_- - a)\mathbb{1}_{\{Y_-\geq a\}} + \mathbb{1}_{\{Y_-<a+1\}} - \mathbb{1}_{\{Y_-<a\}}\right)\Delta N\\ &= \sigma S_- \varphi \Delta M = \varphi \Delta S.\end{aligned}$$

Since m and S are purely discontinuous martingales, we deduce that $m - m_0 = \varphi \cdot S$. Therefore, the proposition follows from Theorem 5.42. □

5.8.2 NUPBR and Deflators

In this subsection, we study another kind of arbitrage in progressive enlargement, namely NUPBR (see Sect. 1.4 for the definition).

In the first step, we assume condition **(C)**.

Proposition 5.44 *Assume that condition* **(C)** *holds. Define a positive $\mathbb{G}$-local martingale by $L := \mathscr{E}(-\frac{1}{Z}\mathbb{1}_{[\![0,\tau]\!]} \cdot \widehat{m})$, where $\widehat{m}$ is the $\mathbb{G}$-martingale part of m^τ given in* (5.1). *Then, for any $\mathbb{F}$-local martingale X, the stopped martingale X^τ admits L as a $\mathbb{G}$-local martingale deflator, thus NUPBR holds for $(X^\tau, \mathbb{G})$.*

Proof We show that for any $\mathbb{F}$-local martingale X, the process $X^\tau L$ is a $\mathbb{G}$-local martingale. Recall that, by Theorem 5.1, the process

$$\widehat{X}_t := X^\tau_t - \int_0^{t\wedge\tau} \frac{1}{Z_{s-}} d\langle X, m\rangle^{\mathbb{F}}_s = X^\tau_t - \int_0^{t\wedge\tau} \frac{1}{Z_s} d\langle X, m\rangle^{\mathbb{F}}_s$$

is a $\mathbb{G}$-local martingale. By integration by parts and using the fact that the predictable covariation process of a continuous martingale is continuous and does not depend on the filtration, we obtain that

$$LX^\tau = L \cdot X^\tau + X \cdot L + \langle L, X^\tau\rangle^{\mathbb{G}} \overset{\text{mart}}{=} 1\!\!1_{[\![0,\tau]\!]} \frac{L}{Z} \cdot \langle X, m\rangle^{\mathbb{F}} - 1\!\!1_{[\![0,\tau]\!]} \frac{L}{Z} \cdot \langle X, \widehat{m}\rangle^{\mathbb{G}} = 0 \tag{5.19}$$

where $U \overset{\text{mart}}{=} V$ is a notation for $U - V$ is a $\mathbb{G}$-local martingale. □

We now assume that the reference filtration $\mathbb{F}$ is quasi-left continuous meaning that for each $\mathbb{F}$-predictable stopping time T, $\mathcal{F}_T = \mathcal{F}_{T-}$.

We apply the decomposition given in (5.2), keeping the same notation, in particular $\bar{m}$ is the $\mathbb{G}$-martingale associated with m and $\widetilde{R}$ is an $\mathbb{F}$-stopping time $\widetilde{R} := R_{\{\widetilde{Z}_R = 0 < Z_{R-}\}}$, where $R := \inf\{t\,:\, Z_t = 0\}$.

Proposition 5.45 *Let $\mathbb{F}$ be a quasi-left continuous filtration. Define a positive $\mathbb{G}$-local martingale by $L := \mathcal{E}(-\frac{1}{Z_-} 1\!\!1_{[\![0,\tau]\!]} \cdot \bar{m})$, where $\bar{m}$ is the $\mathbb{G}$-local martingale part of m^τ given in* (5.2)*. Then, for any $\mathbb{F}$-local martingale X such that $\Delta X_{\widetilde{R}} = 0$ on $\{\widetilde{R} < \infty\}$, the stopped process X^τ admits L as a $\mathbb{G}$-local martingale deflator, thus NUPBR holds for $(X^\tau, \mathbb{G})$.*

Proof First of all, note that L is positive since

$$\Delta\left(-\frac{1}{Z_-} 1\!\!1_{[\![0,\tau]\!]} \cdot \bar{m}\right) = -\frac{\Delta m}{\widetilde{Z}} 1\!\!1_{[\![0,\tau]\!]} + {}^{p,\mathbb{F}}(1\!\!1_{\{\widetilde{Z}=0<Z_-\}}) 1\!\!1_{[\![0,\tau]\!]} > -1 \tag{5.20}$$

where we used the form of $\bar{m}$ and the relationship $m = \widetilde{Z} - Z_-$ (see Proposition 1.46 (d)). Using integration by parts and the optional decomposition given in Theorem 5.2 for X and then for m, we obtain:

$$\begin{aligned}
LX^\tau &= X^\tau_- \cdot L + L_- \cdot X^\tau + [L, X^\tau] \\
&= X^\tau_- \cdot L + L_- \cdot \bar{X} + \frac{L_-}{\widetilde{Z}} 1\!\!1_{[\![0,\tau]\!]} \cdot [m, X] - L_- 1\!\!1_{[\![0,\tau]\!]} \cdot (\Delta X_{\widetilde{R}} 1\!\!1_{[\![\widetilde{R},\infty[\![})^{p,\mathbb{F}} - 1\!\!1_{[\![0,\tau]\!]} \frac{L_-}{Z_-} \cdot [\bar{m}, X] \\
&= X^\tau_- \cdot L + L_- \cdot \bar{X} + L_- \frac{1}{\widetilde{Z}} 1\!\!1_{[\![0,\tau]\!]} \cdot [\bar{m}, X] + L_- \frac{1}{\widetilde{Z}^2} 1\!\!1_{[\![0,\tau]\!]} \cdot [[m], X] \\
&\quad - L_- \frac{1}{\widetilde{Z}} 1\!\!1_{[\![0,\tau]\!]} \cdot [(\Delta m_{\widetilde{R}} 1\!\!1_{[\![\widetilde{R},\infty[\![})^{p,\mathbb{F}}, X] - L_- 1\!\!1_{[\![0,\tau]\!]} \cdot (\Delta X_{\widetilde{R}} 1\!\!1_{[\![\widetilde{R},\infty[\![})^{p,\mathbb{F}} - \frac{L_-}{Z_-} 1\!\!1_{[\![0,\tau]\!]} \cdot [\bar{m}, X] \\
&=: I_1 + I_2 + I_3 + I_4 + I_5 + I_6 + I_7.
\end{aligned}$$

We look closer at the sum of the third and seventh terms of the last expression,

$$
\begin{aligned}
I_3 + I_7 &= L_- \left(\frac{1}{\widetilde{Z}} - \frac{1}{Z_-} \right) \mathbb{1}_{[\![0,\tau]\!]} \cdot [\bar{m}, X] = -L_- \frac{\Delta m}{\widetilde{Z} Z_-} \mathbb{1}_{[\![0,\tau]\!]} \cdot [\bar{m}, X] \\
&= -\sum L_- \frac{\Delta m}{\widetilde{Z} Z_-} \mathbb{1}_{[\![0,\tau]\!]} \Delta \bar{m} \Delta X,
\end{aligned}
$$

where the third equality comes from the fact that $\{\Delta m \neq 0\}$ is a thin set. We add the term I_4 to the previous two,

$$
\begin{aligned}
I_4 + (I_3 + I_7) &= \sum L_- \frac{1}{\widetilde{Z}^2} \mathbb{1}_{[\![0,\tau]\!]} (\Delta m)^2 \Delta X - \sum L_- \frac{\Delta m}{\widetilde{Z} Z_-} \mathbb{1}_{[\![0,\tau]\!]} \Delta \bar{m} \Delta X \\
&= -\sum L_- \frac{\Delta m}{\widetilde{Z}} \Delta X \mathbb{1}_{[\![0,\tau]\!]} \left(\frac{1}{Z_-} \Delta \bar{m} - \frac{1}{\widetilde{Z}} \Delta m \right) \\
&= \sum L_- \frac{\Delta m}{\widetilde{Z}} \Delta X \mathbb{1}_{[\![0,\tau]\!]} \, {}^{p,\mathbb{F}} \left(\mathbb{1}_{[\![\widetilde{R}]\!]} \right),
\end{aligned}
$$

where the last equality comes from (5.20). Note that, by Proposition 1.16 (c) and properties of dual predictable projection, the fifth term in the expression for LX^τ is equal to

$$
\begin{aligned}
I_5 &= -L_- \frac{1}{\widetilde{Z}} \mathbb{1}_{[\![0,\tau]\!]} \cdot [(\Delta m_{\widetilde{R}} \mathbb{1}_{[\![\widetilde{R},\infty[\![})^{p,\mathbb{F}}, X] = -L_- \frac{1}{\widetilde{Z}} \mathbb{1}_{[\![0,\tau]\!]} \, {}^{p,\mathbb{F}} (\Delta m_{\widetilde{R}} \mathbb{1}_{[\![\widetilde{R}]\!]}) \cdot X \\
&= \sum L_- \frac{Z_-}{\widetilde{Z}} \mathbb{1}_{[\![0,\tau]\!]} \, {}^{p,\mathbb{F}} (\mathbb{1}_{[\![\widetilde{R}]\!]}) \Delta X.
\end{aligned}
$$

Finally, using the fact that $Z_- + \Delta m = \widetilde{Z}$, we get

$$
\begin{aligned}
I_5 + (I_4 + I_3 + I_7) &= \sum L_- \frac{Z_-}{\widetilde{Z}} \mathbb{1}_{[\![0,\tau]\!]} \, {}^{p,\mathbb{F}} (\mathbb{1}_{[\![\widetilde{R}]\!]}) \Delta X + \sum L_- \frac{\Delta m}{\widetilde{Z}} \Delta X \mathbb{1}_{[\![0,\tau]\!]} \, {}^{p,\mathbb{F}} \left(\mathbb{1}_{[\![\widetilde{R}]\!]} \right) \\
&= \sum L_- \mathbb{1}_{[\![0,\tau]\!]} \, {}^{p,\mathbb{F}} (\mathbb{1}_{[\![\widetilde{R}]\!]}) \Delta X.
\end{aligned}
$$

Summing up we have that

$$
LX^\tau = X^\tau_- \cdot L + L_- \cdot \bar{X} + \sum L_- \mathbb{1}_{[\![0,\tau]\!]} \, {}^{p,\mathbb{F}} (\mathbb{1}_{[\![\widetilde{R}]\!]}) \Delta X - L_- \mathbb{1}_{[\![0,\tau]\!]} \cdot \left(\Delta X_{\widetilde{R}} \mathbb{1}_{[\![\widetilde{R},\infty[\![} \right)^{p,\mathbb{F}} .
$$

Since X is an $\mathbb{F}$-quasi-left continuous martingale and $\Delta X_{\widetilde{R}} = 0$ on $\{\widetilde{R} < \infty\}$, we see that L is an $\mathbb{G}$-local martingale deflator for X^τ. □

The next result provides an equivalent condition for NUPBR stability.

Theorem 5.46 *The following conditions are equivalent.*
(a) The $\mathbb{F}$-stopping time $\widetilde{R}$ is infinite ($\widetilde{R} = \infty$).
(b) For any $\mathbb{F}$-local martingale X, the process X^τ admits a $\mathbb{G}$-local martingale deflator (hence, satisfies NUPBR($\mathbb{G}$)).

Proof We only give a proof when $\mathbb{F}$ is a quasi-left continuous filtration. For the full proof we refer to [2, 10]. The implication (a) ⇒(b) follows from Proposition 5.45. To prove (b) ⇒(a), we consider the $\mathbb{F}$-martingale

$$X = 1\!\!1_{[\![\widetilde{R},\infty[\![} - \left(1\!\!1_{[\![\widetilde{R},\infty[\![}\right)^{p,\mathbb{F}}.$$

Note that $\mathbb{P}(\tau = \widetilde{R}) = \mathbb{E}\left[\Delta A^o_{\widetilde{R}}\right] = \mathbb{E}\left[\widetilde{Z}_{\widetilde{R}} - Z_{\widetilde{R}}\right] = 0$ (since $0 = \widetilde{Z}_{\widetilde{R}} \geq Z_{\widetilde{R}} \geq 0$). This implies that $\tau < \widetilde{R}$ and

$$X^\tau = -\left(1\!\!1_{[\![\widetilde{R},\infty[\![}\right)^{p,\mathbb{F}}_{\cdot\wedge\tau}$$

is a predictable decreasing process. Thus, X^τ satisfies NUPBR($\mathbb{G}$) if and only if it is a null process. Then, we conclude that $\widetilde{R}$ is infinite. □

The study of NUPBR after τ is more delicate. The following result (among others) is established in [6].

Proposition 5.47 *Let τ be an $\mathbb{F}$-honest time that such that $Z_\tau < 1$ and let $\mathbb{F}$ be a quasi-left-continuous filtration. Then the following assertions are equivalent.*
(a) *The set $\{\widetilde{Z} = 1 > Z_-\}$ is evanescent.*
(b) *For every (bounded) X satisfying NUPBR($\mathbb{F}$), $X - X^\tau$ satisfies NUPBR($\mathbb{G}$).*

5.9 Applications of $\mathscr{E}$-Times to Finance

We have seen in Sect. 4.5, that for an $\mathscr{E}$-time, a specific change of probability measure leads to independence. We exhibit here another interesting change of probability measure. We prove that, for an $\mathscr{E}$-time τ and a finite fixed T, in order to compute conditional expectation of claims of the form $X1\!\!1_{\{T<\tau\}}$ where $X \in \mathscr{F}_T$, or $H_\tau 1\!\!1_{\{\tau<T\}}$ for an $\mathbb{F}$-predictable process H, one can always assume that immersion holds under the pricing measure. We assume null interest rate.

Assume that, under the pricing measure $\mathbb{P}$, the Azéma supermartingale admits a multiplicative decomposition of the form $Z = Ne^{-\Lambda}$ where $\Lambda_t = \int_0^t \lambda_u \eta(du)$ and η is the law of τ, and N is an $\mathbb{F}$-martingale. Then, by Lemma 2.9, the risk-neutral price under $\mathbb{P}$, for every $t \in [0, T]$, equals

$$\mathbb{E}_\mathbb{P}\left[X1\!\!1_{\{\tau>T\}} \,|\, \mathscr{G}_t\right] = 1\!\!1_{\{\tau>t\}} \frac{\mathbb{E}_\mathbb{P}\left[XZ_T \,|\, \mathscr{F}_t\right]}{Z_t} = 1\!\!1_{\{\tau>t\}} \frac{\mathbb{E}_\mathbb{P}\left[XN_T e^{-\Lambda_T} \,|\, \mathscr{F}_t\right]}{N_t e^{-\Lambda_t}}.$$

Under hypothesis $\mathscr{E}$, $\mathbb{P}(\tau > \theta|\mathscr{F}_t) = \int_\theta^\infty p_t(u)\eta(du)$. Let $\mathbb{Q}$ be defined on $\mathscr{G}_T$ as $d\mathbb{Q}|_{\mathscr{G}_T} = L_T d\mathbb{P}|_{\mathscr{G}_T}$, where L is the $(\mathbb{G}, \mathbb{P})$-martingale

$$L_t = 1\!\!1_{\{t<\tau\}} + 1\!\!1_{\{t\geq\tau\}} \lambda_\tau e^{-\Lambda_\tau} \frac{N_t}{p_t(\tau)}.$$

Note that $\mathbb{Q}$ and $\mathbb{P}$ coincide on $\mathcal{G}_\tau$ and that $d\mathbb{Q}|_{\mathcal{F}_t} = N_t d\mathbb{P}|_{\mathcal{F}_t}$. Indeed,

$$\begin{aligned}\mathbb{E}_{\mathbb{P}}[L_t|\mathcal{F}_t] &= Z_t + \int_0^t \lambda_u e^{-\Lambda_u} \frac{N_t}{p_t(u)} p_t(u)\eta(du) \\ &= N_t e^{-\Lambda_t} + N_t \int_0^t \lambda_u e^{-\Lambda_u}\eta(du) = N_t e^{-\Lambda_t} + N_t(1 - e^{-\Lambda_t}) = N_t \,.\end{aligned}$$

Then, for $t \geq \theta$,

$$\begin{aligned}\mathbb{Q}(\tau > \theta|\mathcal{F}_t) &= \frac{1}{N_t}\mathbb{E}_{\mathbb{P}}[L_t \mathbb{1}_{\{\theta<\tau\}}|\mathcal{F}_t] = \frac{1}{N_t}\mathbb{E}_{\mathbb{P}}\left[\mathbb{1}_{\{t<\tau\}} + \mathbb{1}_{\{t\geq\tau>\theta\}}\lambda_\tau e^{-\Lambda_\tau}\frac{N_t}{p_t(\tau)}\Big|\mathcal{F}_t\right] \\ &= \frac{1}{N_t}\left(N_t e^{-\Lambda_t} + \int_\theta^t \lambda_u e^{-\Lambda_u}\frac{N_t}{p_t(u)} p_t(u)\eta(du)\right) \\ &= \frac{1}{N_t}\left(N_t e^{-\Lambda_t} + N_t(e^{-\Lambda_\theta} - e^{-\Lambda_t})\right) = e^{-\Lambda_\theta}\end{aligned}$$

which proves that $\mathbb{Q}(\tau > \theta|\mathcal{F}_t) = \mathbb{Q}(\tau > \theta|\mathcal{F}_\theta)$. Then, by Lemma 3.8, the immersion under $\mathbb{Q}$ holds and the compensator of τ is the same under $\mathbb{P}$ and $\mathbb{Q}$. It follows that

$$\mathbb{E}_{\mathbb{P}}[X \mathbb{1}_{\{T<\tau\}}|\mathcal{G}_t] = \mathbb{E}_{\mathbb{Q}}\left[X \mathbb{1}_{\{T<\tau\}}|\mathcal{G}_t\right] = \mathbb{1}_{\{t<\tau\}}\frac{1}{e^{-\Lambda_t}}\mathbb{E}_{\mathbb{Q}}\left[e^{-\Lambda_T}X|\mathcal{F}_t\right].$$

5.10 Bibliographic Notes

The first and the most important papers on progressive enlargement of filtration include Brémaud and Yor [45] (devoted to immersion), Jacod [122, 123], Jeulin and Yor [140, 141], Meyer [177] and Yor [214]. Many results can be found in Dellacherie et al. [75, Chap. XX]. Jeulin and Yor [144] is also a good summary of main results. A non-exhaustive list of references contains the papers of Ankirchner et al. [20], Jeulin [137], Li and Rutkowski [169], Nikeghbali [183], Song [197, 199–201] and Yoeurp [213]. We refer also to various works by Øksendal and co-authors, based on a different methodology, using forward integrals.

The specific case of honest times has been studied in full generality in Barlow [27, 28], Dellacherie and Meyer [78], Jeulin [138, Chap. 5], Jeulin and Yor [140] and in more recent papers (under **(CA)** conditions) as Nikeghbali and Platen [184] and Nikeghbali [182] (where interesting results on path decomposition are presented). Extensions of Theorem 5.16 beyond conditions **(CA)** are obtained in Acciaio and Penner [3] and Kardaras [153]. See also Li and Rutkowski [169] for a definition and results on pseudo-honest times. Thin times are studied in Aksamit et al. [9] and Aksamit [4, Chap. 2 and Sect. 7.4]. The property of pseudo-stopping time was firstly introduced in the paper by Williams [210], and then was systematically developed in Nikeghbali and Yor [185]. See also Coculescu and Nikeghbali [62] and Aksamit and Li [12] for further results on this class of times.

A systematic study of the properties of hypothesis $\mathcal{E}$ is done in Callegaro et al. [48]. The $\mathcal{E}$-times are the main tool to model default times in El Karoui et al. [86, 87]. Progressive enlargement with $\mathcal{J}$-times is studied in Jeanblanc and Le Cam [130]. The case where the random time is the infimum of a thin time and a $\mathcal{J}$-time is presented in Jiao and Li [146, 147] with the application to sovereign debt.

The paper of Mortimer and Williams [179] deals with changes of measure up to a random time. These results were extended in Kreher [162] in order to study arbitrages before τ under **(CA)** conditions.

There are some studies concerning the case where the enlarged filtration $\mathbb{G}$ is the smallest filtration such that a given random time τ is a stopping time and a given random variable Y is $\mathcal{G}_\tau$-measurable: see Meyer [177], Kchia and Protter [155] in the case where the pair (τ, Y) satisfies Jacod's hypothesis and Jiao and Kharroubi [145] for the case where τ is an $\mathbb{F}$ stopping time. Tian et al. [207] present a study of the PRP in that setting, and Callegaro et al. [47] study optimal investment in that framework.

The case where $\mathbb{G} = \mathbb{F} \vee \widetilde{\mathbb{F}}$ is less studied. Chui [59], Corcuera and Valdivia [66] and Kohatsu-Higa and Yamazato [159] present the case where $\widetilde{\mathbb{F}}$ is the natural filtration of a process I of the form $I_t = G(X, Y_t)$ for an $\mathcal{F}_T$-measurable r.v. X. One can quote also Yor [215] for the case where a Brownian filtration is enlarged so that the process $\int_t^\infty B_s d\mu(s)$ is adapted, where μ is a measure on $\mathbb{R}^+$ and Kchia and Protter [155] for a general study. See Song [196, 198] where a general methodology and results are presented. Kchia and Protter [155] study criteria under which hypothesis $\mathcal{H}'$ is satisfied, approximating the filtration $\widetilde{\mathbb{F}}$ by a filtration enlarged with finite number of random variables.

The multiplicative decomposition appears in Kardaras [152] and Nikeghbali and Yor [186] and a general study can be found in Song [202].

Cases similar to Sect. 5.7 can be found in Jeulin [138, Section 6.3] for the case where the future infimum process is $J_t = \inf_{s \geq t} X_s$ when X is a 3-dimensional Bessel process, and in Yor [219, Chap. 12]).

Coincidence of filtrations from Definition 1.28 used in this chapter is related to the studies on the interplay between progressive and initial enlargements Callegaro et al. [48] and Kchia et al. [154].

The predictable representation property is studied in Barlow [26, 28] for honest times and in Kusuoka [163] under immersion, as we have presented in Chap. 2. Blanchet-Scalliet and Jeanblanc [41] work in a general setting under condition **(C)**. Choulli et al. present an optional martingale representation theorem in [57] and Jeanblanc and Song [135] study the predictable representation property. The particular case of a Poisson filtration is presented in Aksamit et al. [11]. See also Calzagori and Torti [49] who study a general enlargement $\mathbb{F}^X \vee \mathbb{F}^Y$ and give conditions such that the triplet $X, Y, [X, Y]$ has the PRP. The martingale characterization result appears in El Karoui et al. [86, Theorem 5.7] and is generalized in a multidefaut setting in [87]. The notion of multiplicity of a filtration can be found in Davis and Varaiya [69] and Duffie [84].

The study of arbitrages in a progressively enlarged filtration is more recent. The first papers dealing with arbitrages related to honest times are Imkeller [117] and Zwierz [222]. They consider the case of arbitrages occurring after τ under condition **(C)**, proving that there does not exist an equivalent martingale measure. The results we presented here originate from Fontana et al. [100] for the case where condition **(C)** is satisfied and Aksamit et al. [6, 7] for the general case. The study for general random times, based on the existence of a local deflator is provided in Acciaio et al. [2] and Aksamit et al. [7, 8, 10].

Successive enlargements of filtration methodology is applied to insider trading in Blanchet-Scalliet et al. [40] and to optimal investment in Callegaro et al. [47].

Yor [217] was interested in the improvements of Burkholder–Davis–Gundy inequalities, while stopping a continuous martingale at a random time. Beiglböck and Siorpaes [31] and Siorpaes [193] established pathwise Burkholder–Davis–Gundy inequalities in discrete and continuous time which are applied to pseudo-stopping times.

5.11 Exercises

Exercise 5.1 Let τ be an honest time and h be a bounded Borel function. Prove that $\mathbb{E}[h(\tau)|\mathscr{F}_t] = h(\tau_t)(1-Z_t) + \mathbb{E}\left[\int_t^\infty h(s)dA_s^p|\mathscr{F}_t\right]$.

Exercise 5.2 Assume $\mathbb{F}$ is a Brownian filtration. Prove that, if Z is continuous and $Z = N/N^*$ is its multiplicative decomposition, then a $\mathbb{G}$-deflator is $L = \frac{1}{N^\tau}$.

Exercise 5.3 Let τ and τ^* be two honest times. Show that $\tau \vee \tau^*$ is an honest time.

Exercise 5.4 Let τ be an $\mathscr{E}$-time. Prove that hypothesis $\mathscr{H}'$ between $\mathbb{G}$ and $\mathbb{F}^{\sigma(\tau)}$ holds, i.e., that any $\mathbb{G}$-martingale is an $\mathbb{F}^{\sigma(\tau)}$-semimartingale.

Exercise 5.5 Show that the $\mathbb{G}$-optional projection of the $\mathbb{F}^{\sigma(\tau)}$-martingale $L = \frac{1}{p(\tau)}$ is $\ell_t = \mathbb{1}_{\{\tau\le t\}}\frac{1}{p_t(\tau)} + \mathbb{1}_{\{\tau>t\}}\frac{1}{Z_t}\mathbb{P}(\tau>t)$.

Exercise 5.6 Prove that Z_τ has a uniform law if τ is a pseudo-stopping time and Z is continuous,.

Exercise 5.7 Let τ be a pseudo-stopping time and let z be an $\mathbb{F}$-predictable process such that $\mathbb{E}[|z_\tau|] < \infty$. Assume condition **(A)** and that Z is positive. Prove that

$$\mathbb{E}[z_\tau|\mathscr{G}_t] = \mu_0 + \int_0^{t\wedge\tau}\frac{d\mu_s}{Z_s^\tau} + \int_0^t (z_s - h_s)dM_s,$$

where $\mu_t = \mathbb{E}[\int_0^\infty z_s dA_s^p|\mathscr{F}_t]$ and $h_t = (Z_t)^{-1}\left(\mu_t - \int_0^t z_s dA_s^p\right)$.

References

1. B. Acciaio, J. Backhoff Veraguas, A. Zalashko, Causal optimal transport and its links to enlargement of filtrations and continuous-time stochastic optimization (2016), arXiv:1611.02610
2. B. Acciaio, C. Fontana, C. Kardaras, Arbitrage of the first kind and filtration enlargements in semimartingale financial models. Stoch. Process. Appl. **126**, 1761–1784 (2016)
3. B. Acciaio, I. Penner, Characterization of max-continuous local martingales vanishing at infinity. Electron. Commun. Probab. **21** (2016)
4. A. Aksamit, Random times, enlargement of filtration and arbitrages. Ph.D. thesis, University of Evry (2014)
5. A. Aksamit, T. Choulli, J. Deng, M. Jeanblanc, Arbitrages in a progressive enlargement setting. Arbitrage, Credit and Informational Risks, Peking University Series in Mathematics **6**, 55–88 (2014)
6. A. Aksamit, T. Choulli, J. Deng, M. Jeanblanc, No-arbitrage under a class of honest times. Financ. Stoch. forthcoming (2017)
7. A. Aksamit, T. Choulli, J. Deng, M. Jeanblanc, No-arbitrage under additional information for thin semimartingale models (2015), arXiv:1505.00997
8. A. Aksamit, T. Choulli, J. Deng, M. Jeanblanc, No-arbitrage up to random horizon for quasi-left-continuous models. Financ. Stoch. **21**, 1103–1139 (2017)
9. A. Aksamit, T. Choulli, M. Jeanblanc, Classification of random times and applications (2016), arXiv:1605.03905
10. A. Aksamit, T. Choulli, M. Jeanblanc, On an optional semimartingale decomposition and the existence of a deflator in an enlarged filtration. Memoriam Marc Yor, Séminaire de Probabilités XLVII **2137**, 187–218 (2015)
11. A. Aksamit, M. Jeanblanc, M. Rutkowski, Predictable representation property for progressive enlargements of a Poisson filtration (2015), arXiv:1512.03992
12. A. Aksamit, L. Li, Projections, pseudo-stopping times and the immersion property. Séminaire de Probabilités **XLVIII**, 187–218 (2016)
13. J. Amendinger, Initial enlargement of filtrations and additional information in financial markets. Ph.D. thesis, Technischen Universität Berlin (1999)
14. J. Amendinger, Martingale representation theorems for initially enlarged filtrations. Stoch. Process. Appl. **89**, 101–116 (2000)
15. J. Amendinger, D. Becherer, M. Schweizer, A monetary value for initial information in portfolio optimization. Financ. Stoch. **7**(1), 29–46 (2003)

A. Aksamit and M. Jeanblanc, *Enlargement of Filtration with Finance in View*, SpringerBriefs in Quantitative Finance,
https://doi.org/10.1007/978-3-319-41255-9

16. J. Amendinger, P. Imkeller, M. Schweizer, Additional logarithmic utility of an insider. Stoch. Process. Appl. **75**, 263–286 (1998)
17. S. Ankirchner, Information and semimartingales. Ph.D. thesis, Humboldt Universität Berlin (2005)
18. S. Ankirchner, On filtration enlargements and purely discontinuous martingales. Stoch. Process. Appl. **8**, 1662–1678 (2008)
19. S. Ankirchner, S. Dereich, P. Imkeller, The Shannon information of filtrations and the additional logarithmic utility of insiders. Ann. Probab. **34**(2), 743–778 (2006)
20. S. Ankirchner, S. Dereich, P. Imkeller, *Enlargement of filtrations and continuous Girsanov-type embeddings, Séminaire de Probabilités XL*, vol. 1899 (Lecture Notes in Mathematics (Springer, Berlin, 2007)
21. S. Ankirchner, J. Zwierz, Initial enlargement of filtrations and entropy of Poisson compensators. J. Theor. Probab. **24**, 93–117 (2011)
22. S. Asmussen, *Ruin Probabilitiès* (World Scientific, Singapore, 2000)
23. T. Aven, A theorem for determining the compensator of a counting process. Scand. J. Stat. **12**, 69–72 (1985)
24. J. Azéma, Quelques applications de la théorie générale des processusI. I. Invent. Math. **18**(3), 293–336 (1972)
25. J. Azéma, M. Yor, Etude d'une martingale remarquable, *Séminaire de Probabilités XXIII*, vol. 1372, Lecture Notes in Mathematics (1989), pp. 88–130
26. M.T. Barlow, Martingale representation with respect to expanded σ-fields. Unpublished (1977)
27. M.T. Barlow, Decomposition of a Markov process at an honest time. Unpublished (1978)
28. M.T. Barlow, Study of filtration expanded to include an honest time. Zeitschrift für Wahrscheinlichkeitstheorie und verwandte Gebiete **44**, 307–323 (1978)
29. F. Baudoin, Modeling anticipations on financial markets, *Paris-Princeton Lecture on Mathematical Finance 2002*, vol. 1814, Lecture Notes in Mathematics (Springer, Berlin, 2003), pp. 43–92
30. M. Bedini, Information on a default time: Brownian bridges on stochastic intervals and enlargement of filtrations. Ph.D. thesis, Jena University (2012)
31. M. Beiglböck, P. Siorpaes, Pathwise versions of the Burkholder-Davis-Gundy inequality. Bernoulli **21**(1), 360–373 (2015)
32. A. Bélanger, S.E. Shreve, D. Wong, A general framework for pricing credit risk. Math. Financ. **14**(3), 317–350 (2004)
33. K. Bichteler, Stochastic integrators. Bull. Am. Math. Soc. **1**(5), 761–765 (1979)
34. K. Bichteler, Stochastic integration and L^p-theory of semimartingales. Ann. Probab. **9**, 49–89 (1981)
35. T.R. Bielecki, M. Jeanblanc, M. Rutkowski, in *Stochastic processes and applications to mathematical finance, Proceedings of the 5th Ritsumeikan International conference*, ed by J. Akahori, S. Ogawa, S. Watanabe. Hedging of credit derivatives in models with totally unexpected default. (World Scientific, Singapore 2006), pp. 35–100
36. T.R. Bielecki, M. Jeanblanc, M. Rutkowski, Stochastic methods in credit risk modelling, valuation and hedging, in *CIME-EMS Summer School on Stochastic Methods in Finance, Bressanone*, vol. 1856, Lecture Notes in Mathematics, ed. by M. Frittelli, W. Rungaldier (Springer, Berlin, 2004)
37. T.R. Bielecki, M. Jeanblanc, M. Rutkowski, Completeness of a reduced-form credit risk model with discontinuous asset prices. Stoch. Models **22**, 661–687 (2006)
38. T.R. Bielecki, M. Jeanblanc, M. Rutkowski, *Credit Risk Modeling, CSFI Lecture Note Series* (Osaka University Press, Osaka, 2009)
39. T.R. Bielecki, M. Rutkowski, *Credit Risk: Modelling Valuation and Hedging* (Springer, Berlin, 2001)
40. Ch. Blanchet-Scalliet, C. Hillairet, Y. Jiao, Successive enlargement of filtrations and application to insider information. Adv. Appl. Probab. **49**(3), 653–685 (2017)

41. Ch. Blanchet-Scalliet, M. Jeanblanc, Hazard rate for credit risk and hedging defaultable contingent claims. Financ. Stoch. **8**, 145–159 (2004)
42. Ch. Blanchet-Scalliet, F. Patras, *Counterparty risk valuation for CDS, Credit Risk Frontiers: Subprime Crisis, Pricing and Hedging, CVA, MBS, Ratings, and Liquidity* (Wiley, New York, 2009), pp. 437–456
43. V.I. Bogachev, *Measure Theory*, vol. 1 (Springer Science & Business Media, New York, 2007)
44. P. Brémaud, *Point Processes and Queues: Martingale Dynamics* (Springer, Berlin, 1981)
45. P. Brémaud, M. Yor, Changes of filtration and of probability measures. Zeitschrift für Wahrscheinlichkeitstheorie und verwandte Gebiete **45**, 269–295 (1978)
46. D.C. Brody, L.P. Hughston, A. Macrina, Information-based asset pricing. Int. J. Theor. Appl. Financ. **11**, 107–142 (2008)
47. G. Callegaro, M. Gaigi, S. Scotti, C. Sgarra, Optimal investment in markets with over and under-reaction to information. Math. Financ. Econ. **11**(3), 299–322 (2016)
48. G. Callegaro, M. Jeanblanc, B. Zargari, Carthagian enlargement of filtrations. ESAIM: Probab. Stat. **17**, 550–566 (2013)
49. A. Calzolari, B. Torti, Enlargement of filtration and predictable representation property for semi-martingales. Stoch. Int. J. Probab. Stoch. Process. **88**(5), 680–698 (2016)
50. L. Campi, U. Çetin, A. Danilova, Dynamic Markov bridges motivated by models of insider trading. Stoch. Process. Appl. **121**(3), 534–567 (2011)
51. L. Campi, U. Çetin, A. Danilova, Equilibrium model with default and dynamic insider information. Financ. Stoch. **17**(3), 565–585 (2013)
52. L. Campi, U. Cetin, A. Danilova, Explicit construction of a dynamic Bessel bridge of dimension 3. Electron. J. Probab. **18** (2013)
53. R. Carmona, F. Delarue, D. Lacker, Mean field games of timing and models for bank runs. Appl. Math. Opt. **76**(1), 217–260 (Springer, 2017)
54. M. Chaleyat-Maurel, Th. Jeulin, Grossissement gaussien de la filtration brownienne, *Grossissements de filtrations: exemples et applications*, vol. 1118, Lecture Notes in Mathematics (1985), pp. 59–109
55. H.N. Chau, W. Runggaldier, P. Tankov, Arbitrage and utility maximization in market models with an insider (2016), arXiv:1608.02068
56. C.S. Chou, P.-A. Meyer, Sur la représentation des martingales comme intégrales stochastiques dans les processus ponctuels, *Séminaire de Probabilités IX*, vol. 1557, Lecture Notes in Mathematics (Springer, Berlin, 1975), pp. 226–236
57. T. Choulli, C. Daveloose, M. Vanmaele, Hedging mortality risk and optional martingale representation theorem for enlarged filtration (2015), arXiv:1510.05858
58. T. Choulli, J. Deng, Structure condition under initial enlargement of filtration. Sci. China Math. **60**(2), 301–316 (2017)
59. C.K. Chui, The mathematics of insider trading. Ph.D. thesis, University of Capetown (2008)
60. D. Coculescu, From the decompositions of a stopping time to risk premium decompositions. ESAIM: Proc. Surv. **60**, 1–20
61. D. Coculescu, M. Jeanblanc, A. Nikeghbali, Default times, non arbitrage conditions and change of probability measures. Financ. Stoch. **16**(3), 513–535 (2012)
62. D. Coculescu, A. Nikeghbali, Hazard processes and martingale hazard processes. Math. Financ. **3**, 519–537 (2012)
63. S. Cohen, R.J. Elliott, *Stochastic Calculus and Applications*, 2nd edn. (Birkhäuser, Basel, 2015)
64. P. Collin-Dufresne, R. Goldstein, J. Hugonnier, A general formula for valuing defaultable securities. Econometrica **72**, 1377–1407 (2004)
65. J.M. Corcuera, P. Imkeller, A. Kohatsu-Higa, D. Nualart, Additional utility of insiders with imperfect dynamical information. Financ. Stoch. **8**, 437–450 (2004)
66. J.M. Corcuera, V. Valdivia, Enlargements of filtrations and applications (2012), arXiv:1201.5870
67. S. Crépey, M. Jeanblanc, D. Wu, Informationally dynamized Gaussian copula. IJTAF **16** (2013)

68. A. Danilova, M. Monoyios, A. Ng, Optimal investment with inside information and parameter uncertainty. Math. Financ. Econ. **3**(1), 13–38 (2010)
69. M.H.A. Davis, P. Varaiya, On the multiplicity of an increasing family of sigma-fields. Ann. Probab. **2**, 958–963 (1974)
70. F. Delbaen, W. Schachermayer, *The Mathematics of Arbitrage* (Springer, Berlin, 2005)
71. C. Dellacherie, Un exemple de la théorie générale des processus, *Séminaire de Probabilités IV*, vol. 124, Lecture Notes in Mathematics (Springer, Berlin, 1970), pp. 60–70
72. C. Dellacherie, *Capacités et processus stochastiques, Ergebnisse*, vol. 67 (Springer, Berlin, 1972)
73. C. Dellacherie, Un survol de la théorie de l'intégrale stochastique. Stoch. Process. Appl. **10**(2), 115–144 (1980)
74. C. Dellacherie, C. Doléans-Dade, Un contre exemple au problème de Laplacien approché, *Séminaire de Probabilités V*, vol. 191, Lecture Notes in Mathematics (Springer, Berlin, 1971), pp. 127–137
75. C. Dellacherie, B. Maisonneuve, P.-A. Meyer, *Probabilités et Potentiel, chapitres XVII-XXIV, Processus de Markov (fin)* (Compléments de calcul stochastique (Hermann, Paris, 1992)
76. C. Dellacherie, P.-A. Meyer, *Probabilités et potentiel: Chapitres 1 à 4 entièrement refondus* (Hermann, Paris, 1975)
77. C. Dellacherie, P.-A. Meyer, *Probabilités et Potentiel, chapitres I-IV* (Hermann, Paris, 1975). English translation: Probabilities and Potentiel, chapters I-IV (North-Holland, 1982)
78. C. Dellacherie, P.-A. Meyer, A propos du travail de Yor sur les grossissements des tribus, *Séminaire de Probabilités XII*, vol. 649, Lecture Notes in Mathematics (Springer, Berlin, 1978), pp. 69–78
79. C. Dellacherie, P.-A. Meyer, *Probabilités et Potentiel, chapitres V-VIII* (Hermann, Paris, 1980). English translation: Probabilities and Potentiel, chapters V-VIII (North-Holland, 1982)
80. G. Di Nunno, T. Meyer-Brandis, B. Øksendal, F. Proske, Optimal portfolio for an insider in a market driven by Lévy processes. Quant. Financ. **6**, 83–94 (2006)
81. P. Di Tella, H.-J. Engelbert, The predictable representation property of compensated-covariation stable families of martingales. Theory Probab. Appl. **60**, 99–130 (2015)
82. P. Di Tella, H.-J. Engelbert, Martingale Representation, BSDE's and Logarithmic Utility Maximization in Progressively Enlarged Lévy Filtrations, Private communication (2017)
83. O. Draouil, B. Øksendal, A Donsker delta functional approach to optimal insider control and applications to finance. Commun. Math. Stat. **3**(3), 365–421 (2015)
84. D. Duffie, Stochastic equilibria: existence, spanning number, and the no expected financial gain from trade' hypothesis. Econom.: J. Econom. Soc. **54**, 1161–1183 (1986)
85. P. Ehlers, Ph Schönbucher, Background filtrations and canonical loss processes for top-down models of portfolio credit risk. Financ. Stoch. **13**, 79–103 (2009)
86. N. El Karoui, M. Jeanblanc, Y. Jiao, What happens after a default: the conditional density approach. Stoch. Process. Appl. **120**(7), 1011–1032 (2010)
87. N. El Karoui, M. Jeanblanc, Y. Jiao, *Dynamics of multivariate default system in random environment* (Stoch. Process, Appl, 2017)
88. R.J. Elliott, M. Jeanblanc, Incomplete markets and informed agents. Math. Method Oper. Res. **50**, 475–492 (1998)
89. R.J. Elliott, M. Jeanblanc, M. Yor, On models of default risk. Math. Financ. **10**, 179–196 (2000)
90. M. Emery, Espaces probabilisés filtrés: de la théorie de Vershik au mouvement Brownien, via des idées de Tsirelson, *Séminaire Bourbaki, 53ième année*, vol. 282, Astérisque (2002), pp. 63–83
91. M. Emery, E. Perkins, La filtration de B+ L. Zeitschrift für Wahrscheinlichkeitstheorie und verwandte Gebiete **59**(3), 383–390 (1982)
92. P. Ernst, L.C.G. Rogers, Q. Zhou, The value of foresight. Stoch. Process. Appl. forthcoming (2017)
93. S.N. Ethier, ThG Kurtz, *Markov Processes: Characterization and Convergence*, vol. 282 (Wiley, New york, 2009)

94. J.-P. Florens, D. Fougère, Noncausality in continuous time. Econometrica **64**, 1195–1212 (1996)
95. H. Föllmer, P. Imkeller, Anticipation cancelled by a Girsanov transformation: a paradox on Wiener space. Ann. Inst. H. Poincaré Probab. Stat. **26**, 569–586 (1993)
96. H. Föllmer, P. Protter, Local martingales and filtration shrinkage. ESAIM: Probab. Stat. **15**, S25–S38 (2011)
97. H. Föllmer, C.-T. Wu, M. Yor, Canonical decomposition of linear transformations of two independent Brownian motions motivated by models of insider trading. Stoch. Process. Appl. **84**, 137–164 (1999)
98. C. Fontana, Weak and strong no-arbitrage conditions for continuous financial markets. IJTAF **18** (2015)
99. C. Fontana, The strong predictable representation property in initially enlarged filtrations under the density hypothesis. Stoch. Process. Appl. forthcoming (2017)
100. C. Fontana, M. Jeanblanc, S. Song, On arbitrages arising with honest times. Financ. Stoch. **18**, 515–543 (2014)
101. J.-P. Fouque, G. Papanicolaou, R. Sircar, *Perturbation analysis for investment portfolios under partial information with expert opinions* (Soc. Sci. Res. Netw. Work. Pap, Ser, 2014)
102. L. Galtchouk, Semimartingales of processes with independent increments and enlargement of filtration. Theory Probab. Appl. **38**, 395–404 (1994)
103. P. Gapeev, M. Jeanblanc, L. Li, M. Rutkowski, Constructing random times with given survival process and applications to valuation of credit derivatives, in *Contemporary Quantitative Finance*, ed. by C. Chiarella, A. Novikov (Springer, Berlin, 2010)
104. D. Gasbarra, E. Valkeika, L. Vostrikova, Enlargement of filtration and additional information in pricing models: a Bayesian approach, in *From Stochastic Calculus to Mathematical Finance: The Shiryaev Festschrift*, ed. by Yu. Kabanov, R. Lipster, J. Stoyanov (2006), pp. 257–286
105. K. Giesecke, Credit risk modeling and valuation: an introduction, in *Credit Risk: Models and Management*, ed. by D. Shimko (RISK Books, 2004), pp. 487–526
106. Ch. Gouriéroux, A. Monfort, J.-P. Renne, Pricing default events: surprise, exogeneity and contagion. J. Econom. **182**, 397–411 (2014)
107. C.W.J. Granger, Investigating causal relations by econometric models and cross-spectral methods. Econometrica **37**, 424–438 (1969)
108. A. Grorud, M. Pontier, Insider trading in a continuous time market model. Int. J. Theor. Appl. Financ. **1**, 331–347 (1998)
109. A. Grorud, M. Pontier, Comment détecter le délit d'initiés? CRAS Paris série **I**(324), 1137–1142 (1999)
110. A. Grorud, M. Pontier, Asymmetrical information and incomplete markets. Int. J. Theor. Appl. Financ. **4**, 285–302 (2001)
111. X. Guo, Y. Zeng, Intensity process and compensator: a new filtration expansion approach and the Jeulin-Yor formula. Ann. Appl. Probab. **18**(1), 120–142 (2008)
112. S.-W. He, J.-G. Wang, J.-A. Yan, *Semimartingale Theory and Stochastic Calculus* (CRC Press Inc, Boca Raton, 1992)
113. M. Herdegen, S. Herrmann, Single jump processes and strict local martingales. Stoch. Process. Appl. **126**(2), 337–359 (2016)
114. C. Hillairet, Comparison of insiders' optimal strategies depending on the type of side-information. Stoch. Process. Appl. **115**, 1603–1627 (2005)
115. C. Hillairet, Y. Jiao, *Portfolio Optimization with Different Information Flow* (ISTE Press, London)
116. C. Hillairet, Y. Jiao, Portfolio optimization with insider's initial information and counterparty risk. Financ. Stoch. **19**(1), 109–134 (2015)
117. P. Imkeller, Random times at which insiders can have free lunches. Stoch. Stoch. Rep. **74**, 465–487 (2002)
118. P. Imkeller, Malliavin's calculus in insider models: additional utility and free lunches. Math. Financ. **13**, 153–169 (2003)

119. P. Imkeller, N. Perkowski, The existence of dominating local martingale measures. Financ. Stoch. **19**(4), 685–717 (2015)
120. P. Imkeller, M. Pontier, F. Weisz, Free lunch and arbitrage possibilities in a financial market model with an insider. Stoch. Process. Appl. **92**, 103–130 (2001)
121. K. Itô, Extension of stochastic integrals, in *Proceedings of International Symposium on SDE* (Kyoto, 1976), pp. 95–109
122. J. Jacod, *Calcul stochastique et Problèmes de martingales*, vol. 714 (Lecture Notes in Mathematics (Springer, Berlin, 1979)
123. J. Jacod, Grossissement initial, hypothèse H' et théorème de Girsanov, *Grossissements de filtrations: exemples et applications*, vol. 1118, Lecture Notes in Mathematics, Séminaire de Calcul Stochastique 1982–1983 (Springer, Berlin, 1987)
124. J. Jacod, P. Protter, Time reversal of Lévy processes. Ann. Probab. **16**, 620–641 (1988)
125. J. Jacod, A.N. Shiryaev, *Limit Theorems for Stochastic Processes*, 2nd edn. (Springer, Berlin, 2003)
126. S. Janson, S. M'baye, P. Protter, Absolutely continuous compensators. Int. J. Theor. Appl. Financ. **14**(03), 335–351 (2011)
127. R.A. Jarrow, D. Lando, S.M. Turnbull, A Markov model for the term structure of credit risk spreads. Rev. Financ. Stud. **10**(2), 481–523 (1997)
128. R.A. Jarrow, S.M. Turnbull, Pricing derivatives on financial securities subject to credit risk. J. Financ. **50**(1), 53–85 (1995)
129. M. Jeanblanc, Y. Le Cam, Immersion property and credit risk modelling. Mod. Trends Math. Financ.: Kabanov Festschr. **119**, 99–131 (2009)
130. M. Jeanblanc, Y. Le Cam, Progressive enlargement of filtration with initial times. Stoch. Process. Appl. **119**, 2523–2543 (2009)
131. M. Jeanblanc, M. Rutkowski, *Modeling default risk: an overview, Mathematical Finance: Theory and Practice, Fudan University, Modern Mathematics Series* (High Education Press, Beijing, 2000), pp. 171–269
132. M. Jeanblanc, M. Rutkowski, *Modeling default risk: mathematical tools, in Fixed Income and Credit Risk Modeling and Management* (New York University, Stern School of Business, Statistics and Operations Research Department, Workshop, 2000)
133. M. Jeanblanc, S. Song, Default times with given survival probability and their $\mathbb{F}$-martingale decomposition formula. Stoch. Process. Appl. **121**, 1389–1410 (2011)
134. M. Jeanblanc, S. Song, Explicit model of default time with given survival probability. Stoch. Process. Appl. **121**, 1678–1704 (2011)
135. M. Jeanblanc, S. Song, Martingale representation property in progressively enlarged filtrations. Stoch. Process. Appl. **125**, 4242–4271 (2015)
136. M. Jeanblanc, M. Yor, M. Chesney, *Martingale Methods for financial Markets* (Springer, Berlin, 2007)
137. Th Jeulin, Comportement des semi-martingales dans un grossissement de filtration. Zeitschrift für Wahrscheinlichkeitstheorie und verwandte Gebiete **52**, 149–182 (1980)
138. Th Jeulin, *Semi-martingales et grossissement de filtration*, vol. 833 (Lecture Notes in Mathematics (Springer, Berlin, 1980)
139. Th. Jeulin, Filtrations, sous-filtrations: propriétés élémentaires, in *Hommage à P.-A. Meyer et J. Neveu*, vol. 236, Astérisque (SMF, 1996), pp. 163–170
140. Th. Jeulin, M. Yor, Grossissement d'une filtration et semi-martingales: formules explicites, in *Séminaire de Probabilités XII*, vol. 649, Lecture Notes in Mathematics (Springer, Berlin, 1978), pp. 78–97
141. Th. Jeulin, M. Yor, Nouveaux résultats sur le grossissement des tribus. Ann. Scient. Ec. Norm. Sup. **11**, 429–443 (1978)
142. Th. Jeulin, M. Yor, Inégalité de Hardy, semimartingales et faux-amis, in *Séminaire de Probabilités XIII*, vol. 721, Lecture Notes in Mathematics (Springer, Berlin, 1979), pp. 332–359
143. M. Th Jeulin, Yor, (eds.), *Grossissements de filtrations: exemples et applications*, vol. 1118 (Lecture Notes in Mathematics (Springer, Berlin, 1985)

144. Th. Jeulin, M. Yor, Recapitulatif, in *Grossissements de filtrations: exemples et applications*, vol. 1118, Lecture Notes in Mathematics (1985), pp. 305–315
145. Y. Jiao, I. Kharroubi, Information uncertainty related to marked random times and optimal investment, arXiv:1607.02743
146. Y. Jiao, S. Li, Generalized density approach in progressive enlargement of filtrations. Electron. J. Probab. **20** (2015)
147. Y. Jiao, S. Li, *Modeling sovereign risks: from a hybrid model to the generalized density approach* (Math, Financ, 2016)
148. Y. Kabanov, C. Kardaras, S. Song, On local martingale deflators and market portfolios. Financ. Stoch. **20**(4), 1097–1108 (2016)
149. I. Karatzas, C. Kardaras, The numéraire portfolio in semimartingale financial models. Financ. Stoch. **11**, 447–493 (2007)
150. I. Karatzas, I. Pikovsky, Anticipative portfolio optimization. Adv. Appl. Probab. **28**, 1095–1122 (1996)
151. C. Kardaras, Market viability via absence of arbitrage of the first kind. Financ. Stoch. **16**, 651–667 (2012)
152. C. Kardaras, On the stochastic behaviour of optional processes up to random times. Ann. Appl. Probab. **25**(2), 429–464 (2015)
153. K. Kardaras, On the characterisation of honest times that avoid all stopping times. Stoch. Process. Appl. **124**, 373–384 (2014)
154. Y. Kchia, M. Larsson, P. Protter, Linking progressive and initial filtration expansions, *Malliavin Calculus and Stochastic Analysis*, vol. 34, Springer Proceedings in Mathematics and Statistics (2013), pp. 469–487
155. Y. Kchia, P. Protter, Progressive filtration expansions via a process, with applications to insider trading. Int. J. Theor. Appl. Financ. **18**, 1550027 (2014)
156. A. Kohatsu-Higa, Enlargement of filtrations and models for insider trading. Stoch. Process. Appl. Math. Financ. (2004), pp. 151–165
157. A. Kohatsu-Higa, A. Sulem, *A Large Trader-insider Model* (World Scientific Publishing, Hackensack, 2006), pp. 101–124
158. A. Kohatsu-Higa, A. Sulem, Utility maximization in an insider influenced market. Math. Financ. **16**(1), 153–179 (2006)
159. A. Kohatsu-Higa, M. Yamazato, Enlargement of filtrations with random times for processes with jumps. Stoch. Process. Appl. **118**(7), 1136–1158 (2008)
160. A. Kohatsu-Higa, M. Yamazato, Insider modelling and logarithmic utility for models with jumps. Appl. Math. Optim. **64**(2), 217–255 (2011)
161. D. Kramkov, W. Schachermayer, The asymptotic elasticity of utility functions and optimal investment in incomplete markets. Ann. Appl. Probab. 904–950 (1999)
162. D. Kreher, Change of measure up to a random time: details. Stoch. Proc. Appl. **127**(5), 1565–1598 (2016)
163. S. Kusuoka, A remark on default risk models. Adv. Math. Econ. **1**, 69–82 (1999)
164. D. Lando, On Cox processes and credit risky securities. Rev. Deriv. Res. **2**, 99–120 (1998)
165. M. Larsson, Filtration shrinkage, strict local martingales and the Föllmer measure. Ann. Appl. Probab. **24**(4), 1739–1766 (2014)
166. G. Last, A. Brandt, *Marked Point Processes on the Real Line. The Dynamic Approach* (Springer, Berlin, 1995)
167. J.A. León, B. Navarro, D. Nualart, An anticipating calculus approach to the utility maximization of an insider. Math. Financ. **13**, 171–185 (2003)
168. L. Li, M. Rutkowski, Random times and multiplicative systems. Stoch. Process. Appl. **122**(55), 2053–2057 (2012)
169. L. Li, M. Rutkowski, Progressive enlargements of filtrations with pseudo-honest times. Ann. Appl. Probab. **24**, 1509–1553 (2014)
170. B. Mallein, M. Yor, Exercices sur les temps locaux de semi-martingales continues et les excursions browniennes (2016), arXiv:1606.07118
171. R. Mansuy, Infinitely divisible in time processes. Ph.D. thesis, Paris VI University (2005)

172. R. Mansuy, M. Yor, Harnesses, Lévy bridges and Monsieur Jourdain. Stoch. Process. Appl. **115**, 329–338 (2005)
173. R. Mansuy, M. Yor, *Random Times and Enlargements of Filtrations in a Brownian Setting*, vol. 1873 (Lectures Notes in Mathematics (Springer, Berlin, 2006)
174. R. Mansuy, M. Yor, *Aspects of Brownian Motion, Universitext* (Springer, Berlin, 2008)
175. G. Mazziotto, J. Szpirglas, Modèle général de filtrage non linéaire et équations différentielles stochastiques associées. Ann. Inst. H. Poincaré **15**, 147–173 (1979)
176. J.-F. Mertens, Théorie des processus stochastiques généraux: applications aux surmartingales. Zeitschrift für Wahrscheinlichkeitstheorie und verwandte Gebiete **22**(1), 45–72 (1972)
177. P.-A. Meyer, Les résultats de Jeulin sur le grossissement des tribus, in *Séminaire de Probabilités XIV*, vol. 191, Lecture Notes in Mathematics (Springer, Berlin, 1980), pp. 173–188
178. M. Monoyios, Optimal investment and hedging under partial and inside information, in *Advanced Financial Modelling*, vol. 8, ed. by H. Albrecher, W.J. Runggaldier, W. Schachermayer (2009), pp. 371–410
179. T.M. Mortimer, D. Williams, Change of measure up to a random time: theory. J. Appl. Probab. **28**, 914–918 (1991)
180. P.A. Mykland, Stable subspaces over regular solutions of martingale problems. University of Bergen, Statistical Report, No. 15 (1986)
181. P.A. Mykland, Statistical causality. University of Bergen, Statistical Report, No. 14 (1986)
182. A. Nikeghbali, Enlargements of filtrations and path decompositions at non stopping times. Probab. Theory Relat. Fields **136**, 524–570 (2006)
183. A. Nikeghbali, An essay on the general theory of stochastic processes. Probab. Surv. **3**, 345–412 (2006)
184. A. Nikeghbali, E. Platen, A reading guide for last passage times with financial applications in view. Financ. Stoch. **17**, 615–640 (2013)
185. A. Nikeghbali, M. Yor, A definition and some properties of pseudo-stopping times. Ann. Probab. **33**, 1804–1824 (2005)
186. A. Nikeghbali, M. Yor, Doob's maximal identity, multiplicative decompositions and enlargements of filtrations, in *Joseph Doob: A Collection of Mathematical Articles in His Memory*, vol. 50, Illinois Journal of Mathematics, ed. by D. Burkholder (2007), pp. 791–814
187. R. Okhrati, A. Balbás, J. Garrido, Hedging of defaultable claims in a structural model using a locally risk-minimizing approach. Stoch. Process. Appl. **124**, 2868–2891 (2014)
188. P. Ouwehand, Enlargement of filtrations – a primer (2007), http://math.sun.ac.za/~peter_ouwehand/Enlargement_Filtrations_Ouwehand.pdf
189. P.E. Protter, *Stochastic Integration and Differential Equations*, 2nd edn. (Springer, Berlin, 2005)
190. D. Revuz, M. Yor, *Continuous Martingales and Brownian Motion*, 3rd edn. (Springer, Berlin, 1999)
191. M. Rindisbacher, Insider information, arbitrage and optimal portfolio and consumption policies. *Arbitrage and Optimal Portfolio and Consumption Policies* (2001)
192. L.C.G. Rogers, D. Williams, *Diffusions, Markov Processes and Martingales, Foundations*, vol. 1, 2nd edn. (Cambridge University Press, Cambridge, 2000)
193. P. Siorpaes, Applications of pathwise Burkholder–Davis–Gundy inequalities (2015), arXiv:1507.01302
194. L. Sitzia, Modelling inefficient markets: three essays. Ph.D. thesis, University of Turin (2011)
195. E.V. Slud, Stability of stochastic integrals under change of filtration. Stoch. Process. Appl. **50**(2), 221–233 (1994)
196. S. Song, Grossissement de filtration et problèmes connexes. Ph.D. thesis, Paris VI University (1997)
197. S. Song, Drift operator in a market affected by the expansion of information flow: a case study (2013), arXiv:1207.1662
198. S. Song, Local martingale deflators for asset processes stopped at a default time S^τ or right before $S^{\tau-}$ (2013), arXiv:1405.4474

199. S. Song, Local solution method for the problem of enlargement of filtration (2013), arXiv:1302.2862
200. S. Song, Dynamic one-default model. Arbitrage, Credit and Informational Risks, Peking University Series in Mathematics **6**, 123–148 (2014)
201. S. Song, Optional splitting formula in a progressively enlarged filtration. ESAIM Probab. Stat. **18**, 829–853 (2014)
202. S. Song, From Doob's maximal identity to Azéma supermartingale (2016), arXiv:1602.04480
203. Ch. Stricker, Quasi-martingales, martingales locales, semimartingales et filtration naturelle. Zeitschrift für Wahrscheinlichkeitstheorie und verwandte Gebiete **39**, 55–63 (1977)
204. Ch. Stricker, Les ralentissements en théorie générale des processus, in *Séminaire de Probabilités XII*, vol. 649, Lecture Notes in Mathematics (Springer, Berlin, 1978), pp. 364–377
205. Ch. Stricker, M. Yor, Calcul stochastique dépendant d'un paramètre. Zeitschrift für Wahrscheinlichkeitstheorie und verwandte Gebiete **45**(2), 109–133 (1978)
206. K. Takaoka, M. Schweizer, A note on the condition of no unbounded profit with bounded risk. Financ. Stoch. **18**(2), 393–405 (2014)
207. K. Tian, D. Xiong, Z. Ye, The martingale representation in a progressive enlargement of a filtration with jumps (2013), arXiv:1301.1119
208. H.V. Weizsacker, Exchanging the order of taking suprema and countable intersections of σ-algebras. Annales de l'institut Henri Poincaré (B) Probabilités et Statistiques. **19**, 91–100 (1983)
209. D. Williams, *Probability with Martingales* (Cambridge University Press, Cambridge, 1991)
210. D. Williams, A non-stopping time with the optional-stopping property. Bull. Lond. Math. Soc. **34**, 610–612 (2002)
211. C.-T. Wu, Construction of Brownian motions in enlarged filtrations and their role in mathematical models of insider trading. Ph.D. thesis, Humbolt University (1999)
212. Ch. Yoeurp, Contributions au calcul stochastique. Ph.D. thesis, Paris VI University (1982)
213. Ch. Yoeurp, *Grossissement de filtration et théorème de Girsanov généralisé, in Grossissements de filtrations: exemples et applications*, vol. 1118 (Lecture Notes in Mathematics (Springer, Berlin, 1985)
214. M. Yor, Grossissement d'une filtration et semi-martingales: théorèmes généraux, in *Séminaire de Probabilités XII*, vol. 649, Lecture Notes in Mathematics (Springer, Berlin, 1978), pp. 61–69
215. M. Yor, Application d'un lemme de Jeulin au grossissement de la filtration brownienne, in *Séminaire de Probabilités XIV*, vol. 784, Lecture Notes in Mathematics (Springer, Berlin, 1980), pp. 189–199
216. M. Yor, Grossissement de filtrations et absolue continuité de noyaux, in *Grossissements de filtrations: exemples et applications*, vol. 1118, Lecture Notes in Mathematics (Springer, Berlin, 1985), pp. 7–14
217. M. Yor, Grossissement progressif: inégalités de martingales continues arrêtées à un temps quelconque. Lect. Notes Math. **1118**, 110–146 (1985)
218. M. Yor, *Some Aspects of Brownian Motion, Part I: Some Special Functionals* (Lectures in Mathematics, ETH Zürich (Birkhäuser, Basel, 1992)
219. M. Yor, *Some Aspects of Brownian Motion, Part II: Some Recent Martingale Problems* (Lectures in Mathematics, ETH Zürich (Birkhäuser, Basel, 1997)
220. Y. Zeng, Compensators of stopping times. Ph.D. thesis, Cornell University (2006)
221. W.-A. Zheng, Une remarque sur une même intégrale stochastique calculée dans deux filtrations, in *Séminaire de Probabilités XVIII*, vol. 1059, Lecture Notes in Mathematics (Springer, Berlin, 1984), pp. 172–173
222. J. Zwierz, On existence of local martingale measures for insiders who can stop at honest times. Bull. Pol. Acad. Sci.: Math. **55**, 183–192 (2007)

Index

A. Aksamit and M. Jeanblanc, *Enlargement of Filtration with Finance in View*, SpringerBriefs in Quantitative Finance,
https://doi.org/10.1007/978-3-319-41255-9

ymbol Index

A. Aksamit and M. Jeanblanc, *Enlargement of Filtration with Finance in View*, SpringerBriefs in Quantitative Finance,
https://doi.org/10.1007/978-3-319-41255-9

Printed in the United States
By Bookmasters